Surge in Solar-Powered Homes

DIRECTIONS IN DEVELOPMENT
Energy and Mining

Surge in Solar-Powered Homes

Experience in Off-Grid Rural Bangladesh

Shahidur R. Khandker, Hussain A. Samad, Zubair K. M. Sadeque,
Mohammed Asaduzzaman, Mohammad Yunus, and A. K. Enamul Haque

 **WORLD BANK GROUP**

Contents

Foreword

Bangladesh has recently been playing a quiet but prominent role in extending the reach of solar power in rural areas. Thanks to the recent surge in solar home system (SHS) installations, millions of people in off-grid areas have electric lighting for the first time, which means children and adults can read and study in the evenings and have a greater sense of nighttime security. With solar-powered television, family members, including women, gain access to knowledge and information, which empowers them and helps them acquire capabilities they did not have earlier. And families discover they can earn income from renting out mobile phone–charging services, helping their neighbors that have not yet purchased an SHS to avoid the cost of frequent commutes to distant phone-charging centers.

This remarkable policy experiment has begun to attract attention, and has great potential not just for Bangladesh but for other developing countries and the advance of clean energy. This, in turn, makes this new book by Shahidur Khandker, Hussain Samad, Zubair Sadeque, Mohammed Asaduzzaman, Mohammad Yunus, and Enamul Haque timely and important.

The benefits from kerosene replacement for lighting alone are enormous. As the book shows, kerosene lighting is more than 35 times as expensive as SHS-powered electric lighting, and SHS owners consume about 3.5 times as much lighting as non-owners. Kerosene replacement with SHS not only offers people a much greater quantity of far-higher-quality lighting, it reduces the health and safety risks linked to kerosene-based lighting, particularly among women and young children. Moreover, nearly 160 million kg in carbon dioxide (CO_2) emissions are avoided each year. Considering that only about 10 percent of people in off-grid areas have adopted SHS to date, the potential for reducing carbon emissions is large.

Currently, many SHS units are being installed in rural Bangladesh under the second phase of the Rural Electrification and Renewable Energy Project, supported by the World Bank and other development partners. The Infrastructure Development Company Limited (IDCOL), the project's implementing agency, has set a target of installing an additional 3 million units within the next two years. A key element of the project's success is IDCOL's innovative, partially subsidized SHS delivery and financing scheme. IDCOL provides participating nongovernmental organizations, called partner organizations, direct incentives that encourage them to lower the SHS unit price to household buyers and microcredit financing, which puts SHS within reach of families who could not otherwise afford the high upfront costs.

Given the rapid growth in SHS, the proven benefits to household adopters, and the future uncertainty of national grid extension, policy makers must ask critical questions at this juncture. They need to know the potential market demand for SHS in off-grid areas. They also need to evaluate whether or to what extent the subsidy should be continued for further SHS expansion in off-grid areas. To answer these questions, they require a better grasp of the nature and quality of the program delivery system, financing mechanism, and market features. This book, which conducts a detailed empirical investigation based on a large-scale household survey and institutional data, is designed to fill that information gap. It is an impressive work, which should be of interest to anyone interested in energy and the environment.

As U.S. Secretary of State and former Senator John Kerry stated in his November 2009 address to the World Bank, the issues of energy poverty and climate change are not mutually exclusive, and Bangladesh's SHS success story suggests the type of innovative projects where the World Bank can make a difference. All in all, this is an important and highly readable book on a topic of great contemporary concern.

Kaushik Basu
Senior Vice President, Development Economics
and Chief Economist
The World Bank

Preface

Provision of electricity is a recognized development agenda of governments and donors, including the World Bank and many countries. Some 1.2 billion people globally still lack access to electricity, which limits their opportunities to improve their welfare. Most of these people are residents of 20 developing countries in Asia and Sub-Saharan Africa, and about 80 percent live in rural areas of those countries. Electricity is, in fact, an integral component of socioeconomic development, with benefits ranging from enhanced income, productivity, and employment resulting from access to electronic media and improved household lighting. Lacking access to electricity is therefore considered a major impediment to growth and development.

But efforts to provide such modern energy as electricity for lighting, heating, cooking, and other production purposes face enormous challenges. Poor households in most countries typically have limited access to, as well as limited ability to pay for, quality energy services. Better access to modern energy means being able to afford and use electricity. Despite national efforts with donor support, expansion of the national electricity grids in many poor countries, such as Bangladesh, is quite slow, mainly because of limited electricity generation and supply resulting from the lack of price and institutional reforms in the power sector.

With increasing technology development via alternative sources of electricity generation, such as solar power, off-grid electrification becomes a viable alternative to conventional electrification approaches. In recent years, a decentralized energy-generation mechanism based on solar photovoltaics has gained currency for promoting solar power as part of achieving universal electrification in the developing world. However, the success of off-grid electrification models has also been limited as it depends critically on consumer demand. Lack of income is a serious bottleneck to adopting solar home systems (SHSs). Poor households must use their limited income to pay for electricity in addition to other essential livelihood items, and thus the spread of SHS has proved challenging.

How can the spread of solar panels and other new technologies be promoted? What could promote SHS adoption, given limited household income and knowledge about solar panel technology? Although the importance of SHS has been recognized for decades, there is a dearth of research on how and when such technology can be promoted.

Bangladesh's experience with SHS has been a phenomenal success, as noted in the energy development literature. Because of its innovative program design, including a price-support scheme to reduce the cost of SHS purchase and maintenance, Bangladesh's success story appears to be an option for accelerating SHS adoption in poor countries. This book offers a snapshot of the history of SHS development in Bangladesh and the program's pioneering role in marketing SHS in a country where most of its clients have limited purchasing power.

What started in 2003 as a five-year project to provide support for reaching 50,000 rural homes in Bangladesh with SHS is now reaching more than 50,000 *every month*, making this the world's fastest-growing SHS program. An impact evaluation study conducted in 2012 has established increased study time for children, increased sense of security, and enhanced women's empowerment as benefits of a well-functioning SHS, which provides both superior lighting quality by replacing traditional kerosene lamps and connectivity to the outside world by powering television and mobile phone chargers.

To date, the SHS program has reached 3 million off-grid rural homes in Bangladesh, with a target of reaching another 3 million in the next couple of years. The phenomenal scale of the program has intrigued the governments of low-access countries, as well as development practitioners from around the world. What contributed to the program's success? How is the quality of the systems ensured? How are the systems made affordable to rural populations with limited income opportunities? This book attempts to address these critical questions.

Bangladesh's SHS program leveraged certain attributes unique to the country that would be difficult to replicate in other countries struggling with the daunting task of increasing rural access to electricity. The program also leveraged other factors that would be easier to replicate. This book strives to explain all of these vital factors. For the non-technical reader, the book explains how a SHS works, its impacts on rural households based on empirical evidence, and the financing mechanism and implementation model that aim to meet rural households' basic electricity needs through a market-based supply mechanism.

Acknowledgments

The authors take this opportunity to recognize the many individuals and institutions that have been instrumental in moving this program forward to where it is today. Special appreciation goes to Fouzul Kabir Khan and Subramaniam V. Iyer for their role in setting the program direction from the outset, which has stood the test of time. Special recognition is extended to the officials and staff of the Infrastructure Development Company Limited (IDCOL), the government-owned financial intermediary and project implementing agency, whose dedication and hard work have been indispensable to the program reaching its current status. The authors are also grateful to all the partner organizations—those doing the difficult day-to-day field work—whose contributions cannot be overemphasized.

Throughout the impact evaluation study and the writing of this book, Mahmood Malik, S. M. Monirul Islam, and Enamul Karim Pavel were steadfast in providing all of the needed support and cooperation. At the World Bank, the authors acknowledge the support of Julia Bucknall, Sector Manager for South Asia Sustainable Development Energy, and Johannes Zutt, Country Director for Bangladesh. Gratitude is extended to the many others who made valuable contributions along this journey. They include Douglas Barnes, Raihan Elahi, Sudeshna Ghosh Banerjee, Luis Andres, Atsushi Limi, Anil Cabraal, Jyoti Shukla, Christine Kimes, Andras Horvai, Luisa Mimmi, Abdul Malek Azad, Sharmind Neelormi, and Md. Amir Hossain. The authors acknowledge the invaluable editorial support provided by Norma Adams and the research assistance of Estella Malayika. They are deeply indebted to William Martin and Raluca Golumbeanu for their ongoing encouragement in bringing this study to fruition. They also thank Stephen McGroarty and Paola Scalabrin in the World Bank Office of the Publisher for their guidance in producing this book.

Finally, the financial support provided by the Global Partnership on Output-Based Aid (GPOBA) is gratefully acknowledged. GPOBA is a global partnership program administered by the World Bank. It was established in 2003 to develop output-based aid (OBA) approaches across a variety of sectors—among them infrastructure, health, and education. To date, GPOBA has signed 38 grant agreements for OBA subsidy funding, for a total of US$155 million. GPOBA projects have disbursed more than US$105 million based on independently verified outputs, directly benefiting more than 7 million people. The program's current donors are the United Kingdom's Department for International Development (DFID), International Finance Corporation (IFC), Directorate-General for International Cooperation of the Dutch Ministry of Foreign Affairs (DGIS), Australian Department of Foreign Affairs and Trade (DFAT), and Swedish International Development Cooperation Agency (Sida).

About the Authors

Shahidur R. Khandker (PhD, McMaster University, Canada, 1983) is a lead economist in the Development Research Group of the World Bank. He has authored more than 40 articles in peer-reviewed journals, including the *Journal of Political Economy, Review of Economic Studies, World Bank Economic Review,* and *Journal of Development Economics.* He has also written various books, including the *Handbook on Impact Evaluation: Quantitative Methods and Practices,* co-authored with Gayatri Koolwal and Hussain Samad and published by the World Bank; *Seasonal Hunger and Public Policies: Evidence from Northwest Bangladesh,* co-authored with Wahiduddin Mahmud; *Fighting Poverty with Microcredit: Experience in Bangladesh,* published by Oxford University Press; and *Handbook on Poverty and Inequality,* co-authored with Jonathan Haughton and published by the World Bank. He has written several book chapters and more than three dozen discussion papers at the World Bank on poverty, rural finance and microfinance, agriculture, and infrastructure. His work spans some 30 countries and covers a wide range of development issues, from microfinance and rural finance, agriculture, and infrastructure to poverty, seasonality, and energy.

Hussain A. Samad (MS, Northeastern University, 1992) is a consultant at the World Bank with more than 18 years of experience in impact evaluation, monitoring and evaluation, data analysis, research, and training on development issues. He has been involved in many World Bank research studies covering such areas as rural electrification, energy poverty, microfinance, poverty, seasonality, migration, and household air pollution. Throughout his long career, he has co-authored books, articles in peer-reviewed journals, and reports; provided technical support to regions, field offices, and consulting firms; presented in seminars; designed course materials for training; and conducted hands-on training in workshop settings.

Zubair K. M. Sadeque (MS, Duquesne University, 2005) is a senior energy specialist at the World Bank, based in Dhaka. He serves as task team leader of the Rural Electrification and Renewable Energy Development Project (RERED II), whose objective is increasing access to clean energy in rural Bangladesh. Concurrently, he manages the Rural Grid Electrification Project in Bangladesh.

Before joining the World Bank in 2008, he was a lecturer in finance at North South University in Dhaka. He earned his CFA charter in 2009.

Mohammed Asaduzzaman (PhD, University of Sussex, 1979) is an economist, currently serving as professorial fellow at the Bangladesh Institute of Development Studies (BIDS), where he was research director. Previously, he was deputy chair of the International Commission on Sustainable Agriculture and Climate Change under the Consultative Group on International Agricultural Research (CGIAR). He is closely involved with the Bangladesh government's climate change management initiatives. He was a major contributor to the Bangladesh Climate Change Strategy and Action Plan (2009). He serves on the boards of the Krishi Gobeshona Foundation and the Krishi Gobeshona Endowment Trust, set up by the Bangladesh government to facilitate funding for agricultural research. His broad research experience covers agriculture, natural resource management, energy, environment and climate change, sustainable development, and rural development. Within these areas, he has worked on technological change in agriculture, coastal environmental management, comprehensive evaluation of poverty eradication programs, rural energy, rural electrification and renewable energy, and local level planning. His recent climate change work has focused on mitigation, economics, mainstreaming in national planning, and energy modeling.

Mohammad Yunus (PhD, Georgia State University, 2006) is a senior research fellow at the Bangladesh Institute of Development Studies (BIDS). His current research interests include taxation, local government fiscal policies, fiscal decentralization, food security and poverty alleviation, and impact evaluation. He has published extensively in both nationally and internationally accredited journals. He also has to his credit numerous other publications in the form of research reports and contribution to various edited volumes.

A. K. Enamul Haque (PhD, University of Guelph, Canada, 1991) is a professor of economics at the East West University in Dhaka. He has served on the management and advisory committees of the South Asian Network for Development and Environmental Economics (SANDEE), which includes Bangladesh, Bhutan, India, Maldives, Nepal, Pakistan, and Sri Lanka. He has published widely on issues related to environment, education, and public policy. In 2011, he edited *Environmental Valuation in South Asia*, published by Cambridge University Press.

Abbreviations

ATT	average treatment of the treated
BIDS	Bangladesh Institute of Development Studies
BUET	Bangladesh University of Engineering and Technology
CFL	compact fluorescent lamp
CO_2	carbon dioxide
COP	Conference of the Parties
CV	contingency valuation
DFID	Department for International Development
FE	fixed-effects (method)
GDP	gross domestic product
GHG	greenhouse gas
GPOBA	Global Partnership on Output-Based Aid
HAP	household air pollution
HIES	Household Income and Expenditure Survey
IDCOL	Infrastructure Development Company Limited
ISO	International Standards Organization
IV	instrumental variables
JICA	Japan International Cooperation Agency
KM	kernel matching
LED	light-emitting diode
LLM	local linear matching
LPG	liquefied petroleum gas
LUTW	Light Up the World
MDG	Millennium Development Goal
MW	megawatt
NGO	nongovernmental organization
NN	nearest neighbor
OBA	output-based aid
OHSAS	Occupational Health and Safety Standards

OLS	ordinary least squares
PO	partner organization
PSM	propensity-score matching
PV	photovoltaic
RDD	regression discontinuity design
REB	Rural Electrification Board
RERED	Rural Electrification and Renewable Energy Development Project
RSF	Rural Services Foundation
SE4ALL	Sustainable Energy for All
SELF	Solar Electric Light Fund
SHS	solar home system
WTP	willingness to pay

CHAPTER 1

Introduction

Providing electricity is a recognized development agenda item and one of the key pillars of the Sustainable Energy for All (SE4ALL) initiative of the United Nations. Some 1.2 billion people globally still lack access to electricity, which limits opportunities to improve their welfare (World Bank 2013). Most of these people reside in 20 developing countries of Asia and Sub-Saharan Africa, and about 80 percent live in rural areas of those countries. Electricity is, in fact, an integral component of socioeconomic development, with myriad benefits, including improved household lighting and access to electronic media, which enhance income, productivity, and employment (Cabraal, Barnes, and Agarwal 2005; Dinkelman 2011; Khandker, Barnes, and Samad 2012). Lacking access to electricity is thus considered a major impediment to development.

Efforts to provide such modern forms of energy as electricity for lighting, heating, cooking, and other productive purposes face enormous challenges. In most countries, poor households typically have limited access to and ability to pay for quality energy services. Better access to modern energy means being able to afford and use electricity. Despite national efforts with donor support, expansion of the national electricity grids in many poor countries, such as Bangladesh, has been quite slow, mainly because of limited electricity generation and supply, resulting from the lack of price and institutional reforms in the power sector (Barnes 2007; Zerriffi 2011).

Challenge of Off-Grid Electrification

With increasing technology development via solar power and other alternative generation sources, off-grid electrification becomes a viable complement to conventional electrification approaches (Brass et al. 2012; Jacobson 2007; Wamukonya 2007; Zerriffi 2011). A decentralized energy-generation mechanism based on solar photovoltaics (PV), for example, has gained prominence in recent years for achieving universal electrification in the developing world. But the success of off-grid models has also been limited since it depends critically on consumer demand. Lack of income is a serious bottleneck to adopting solar home systems (SHSs). Because poor households must allocate their scant incomes to pay for other essential livelihood items, as well as electricity, the spread of SHS has proved challenging (Friebe, von Flotow, and Täube 2013; Nieuwenhout et al. 2001).

Given limited household income and knowledge about solar panel technology, what factors could promote SHS adoption? Although the importance of SHS has been recognized for decades (Nieuwenhout et al. 2001), little research has addressed how and when the spread of solar panels and other new technologies can be promoted. Few studies found in the literature have explored the key determinants of early adoption of SHS when such technologies are made available (Komatsu, Kaneko, and Ghosh 2011; Lay, Ondraczek, and Stoever 2013; Rebane and Barham 2011; Siegel and Rahman 2011). Household income matters a lot, as do cost and technology. Because income is limited in poor countries, any price support toward reducing the cost of SHS purchase and maintenance appears to be an option for accelerating adoption.

Bangladesh's Experience

Bangladesh has made remarkable progress in raising living standards and reducing poverty, particularly in previously lagging regions. From 2005 to 2010, growth in gross domestic product (GDP) reached more than 6 percent a year, and rural poverty fell by 8.5 percent (from 40 percent to 31.5 percent). Yet such positive changes have not been matched by a commensurate rise in energy consumption and access. Peak demand exceeds supply by about 2,000 megawatts (MW) (8,500 versus 6,500 MW). The grid has reached just 42.5 percent of rural households, 12.5 percent less than the national average (BBS 2011). In scattered and remote villages, grid electrification is expensive, which can challenge the financial viability of power utilities. Large industrial loads in urban areas often take priority over the rural countryside, where most Bangladeshis live. Even for rural households with a grid connection, power outages may be frequent and prolonged.[1]

Bangladesh's national strategy calls for achieving universal access to electricity by 2021. Electricity has been a critical input toward achieving the Millennium Development Goals (MDGs), affording households an array of benefits. These range from clean energy for high-quality lighting, which improves health and enables children—both girls and boys—to study for longer periods after sunset, to greater farm- and non-farm productivity, and women's empowerment through better time allocation and access to information. It is unrealistic to expect grid-based electrification alone to result in universal access in the near future. In rural areas that are not economically viable, including newly accreted coastal islands (locally known as *char* lands), off-grid solutions using renewable energy technologies offer a sensible alternative to conventional power supply (Brass et al. 2012; Jacobson 2007; Wamukonya 2007; Zerriffi 2011). Solar PV, in particular, has substantial off-grid potential, given the country's tropical climate, featuring abundant year-round sunshine (Islam, Islam, and Rahman 2006).

Rapid SHS expansion in Bangladesh to some 3 million rural households by early 2014 has drawn the attention of donors and governments of other countries. Phenomenal coverage within such a short period of time has been made possible, in part, by the subsidy provided by donors to facilitate SHS adoption in remote and off-grid areas. Even so, no more than 10 percent of off-grid households have been reached, meaning there is ample scope for continued SHS expansion, particularly given the numerous constraints faced by the current supply of grid-based electricity; these include limited supply, increased dependence on power plants than run on high-cost liquid fuels, and lack of funds.

How to provide people electricity, even in a limited manner, amid increasing demand remains a challenge for policy makers, requiring more information. For example, one must know the reach and effectiveness of the SHS technology in meeting rural people's electricity needs for productive purposes, wherever applicable, and improving their quality of life. One must determine whether the observed impacts of SHS on household welfare recommend its expansion to reach more people in off-grid areas. In addition, one requires a better grasp of the nature and quality of the program delivery system, financing mechanism, and market features. This book is designed to fill that gap.

Study Purpose and Approach

The book's broad aim is twofold: (a) to assess the welfare impact of SHS on households and (b) to evaluate the present institutional structure and financing mechanisms in place, noting that households want cheaper systems and good-quality service while suppliers require a reasonable market-based profit to stay in business. The book's specific objectives are to assess (a) which households in off-grid areas adopt SHS and the direct and indirect benefits for household members, including women and children; (b) the cost-effectiveness of SHS for adopters; (c) the nature and quality of the program delivery system and its differentiation by supplier; (d) market features, including current size and limitations and future potential size in the context of various influencing factors; and (e) role of the financing mechanism, including the effectiveness of subsidies in SHS market expansion, clients' willingness to pay, and resulting household welfare.

The study entailed an intensive empirical investigation based on both primary and secondary data. The primary data consisted mainly of a large-scale, nationally representative household survey with appropriate geographic spread. Conducted in 2012 by the Bangladesh Institute of Development Studies (BIDS) and assisted by the World Bank, the household survey was designed to examine SHS benefits and costs (BIDS/World Bank 2012). In addition, the branch offices of suppliers, known as partner organizations (POs), and local communities were surveyed, using pre-tested questionnaires, to investigate the cost-effectiveness of the SHS technology and delivery system for adopter households. Other primary data included consultations held with key stakeholders, including the Infrastructure Development Company Limited (IDCOL), the implementing agency for the World Bank–supported project, and the various POs.

A total of 4,000 households were surveyed in 128 villages, evenly split between treatment villages (i.e., those with an existing SHS supply) and control villages (i.e., those without a SHS supply). In the treatment villages, 1,600 households had adopted SHS while 400 had not. The control villages comprised 2,000 non-SHS households. The sample of treatment households was randomly selected from a database of nationwide SHS customers maintained by IDCOL (Asaduzzaman et al. 2013).

Several analytical techniques were used to address the study's key objectives. These included a probability function, which was estimated to determine the factors that have played important roles in household adoption of SHS in villages with access to the technology and whether the SHS price, including the subsidy, has affected the adoption rate. Simulation analyses were conducted to estimate the potential future market size for specific SHS capacities and to understand customers' willingness to pay and the implications of various financing mechanisms and subsidy levels.

Research Issues and Key Findings

The book addresses a number of research issues, which are grouped according to general and gendered household impact, program delivery and monitoring of technical standards, market size and demand, and carbon emissions reduction. The book first identifies major factors that determine household access to various SHS capacities and proximate causes of access and types adopted. The major determining variables of access are found to be household wealth, income, landholdings, occupation, and educational level. Rural households that purchase SHS from the POs choose from systems in a capacity range of 20–120 watt-peak (Wp). Three capacity levels (20, 40, and 50 Wp) have dominated the market to date, with 50 Wp models representing the most sales; but 20 Wp systems are quickly gaining in popularity. The major difference among the SHS packages offered is the number of connection points for lighting.

The book also analyzes household uses of solar-electric energy services. Typically, SHS models are used for lighting, powering fans and television sets, and charging mobile devices and other electrical equipment. An immediate benefit of SHS adoption is kerosene substitution for lighting. Replacing smoke-emitting kerosene lamps and lanterns with solar-powered lights offers a far higher quality of lighting without contributing to household air pollution (HAP). These major direct benefits have important implications for the study behavior of school-going children—both girls and boys—and the health of family members, particularly women and young children, who may spend many hours indoors each day.

A major benefit of SHS adoption is better kitchen lighting. Since women spend a major part of their day in the household kitchen or other indoor cooking area, kitchen lighting is a critical issue that affects women's health and time use. Traditionally, women and poorer households have used a one-wick kerosene lamp in the kitchen (locally known as a *kupi*); this primitive lamp emits heavy smoke, with soot causing discoloration of the surrounding walls and ceilings.

The most popular SHSs operate well only within a certain distance of the charge controller—usually located in one of the main rooms of the house—and have only a few connection points, two of which are reserved for the television set and mobile phone charger. This means that kitchens often go without solar-powered lighting, illustrating the importance of the number and location of the connection points and suggesting that replacement of the kupi, versus other types of kerosene lamps, may be more pro-poor and gender-friendly.

Another set of issues examined is the differentiation of SHS impact across households. Since more expensive SHS packages offer more connection points used for lighting, better-off households may enjoy greater kerosene substitution and thus be less exposed to HAP-linked health risks. Key research areas are changes in income and expenditure, including those for energy services and energy-using devices, and their differentials by SHS capacity and socioeconomic status of households.

The book considers how SHS adoption results in changes in time-use patterns of household members. Introducing solar-powered lighting has the immediate effect of extending the waking and working hours of all household members and providing a sense of security. Household chores and productive work, such as sewing, can be done at a less hurried pace, which especially affects women. If the home has solar-powered kitchen lighting, women may spend less time cleaning; food preparation may be done later in the evening at a slower pace, which may affect the nutritional value of food and families' health. Also, more time may be available for evening reading and social interaction.[2] A major shift in remote rural areas resulting from SHS introduction is households' access to television. Providing solar-powered electricity may encourage households to purchase a television set. More leisure time for watching TV and listening to the radio affords household members access to useful information that can improve their health and reproductive behavior, raise awareness of their rights, and offer other positive social values.

SHS adoption also makes it possible for households to generate income from phone chargers. With solar-powered electricity, connected households can offer non-SHS households charger services for mobile phones. Adopter households benefit by earning extra money, while their non-SHS counterparts spend less time and money commuting to more distant charger locations.

Finally, the book evaluates the gender-disaggregated benefits and women's empowerment from SHS adoption. The gender analysis included two major research questions: (a) can the socioeconomic status of rural women be enhanced by increasing their opportunity to participate in alternative energy-service delivery and (b) if SHS brings positive impacts in terms of social indicators, what additional efforts can supplement them to bring about a radical shift in gender roles and responsibilities. Currently, opportunities for women as energy service providers are extremely limited in rural Bangladesh. But women, who represent the largest group of rural energy users, particularly for household cooking, are a natural choice for providing such services. As growth in SHS dissemination continues, it is expected that women's empowerment will increase through technical training in SHS operation.

The book's findings show that better household lighting improves household welfare both directly and indirectly. It increases the study time of school-going boys and girls, which leads to better educational outcomes. Women's decision-making power is facilitated through access to knowledge and information made possible by solar-powered television. However, the upfront cost of acquiring a SHS is high for many rural households. To make the system affordable, IDCOL developed an innovative, partially subsidized delivery scheme within an institutional framework that is quite effective in reaching its clientele base. As the book observes, the subsidy provided by IDCOL's guaranteed refinancing scheme has a positive impact on the price of SHS units. Even though the subsidy is not given directly to household buyers, they still receive a part of it in the form of a lower unit price. Thus, the subsidy indeed trickles down. The analysis further shows that the subsidy has declined over time, from 25 percent of the average unit price in 2004 to 10 percent in 2012. The benefit-cost analysis shows that the social benefits generated for society far exceed the cost of the subsidy.

Clearly, the SHS program's demonstrated success provides evidence for continuing it at the current subsidy rate. Along with well-targeted subsidies, future expansion will require adequate donor financing. Also, the system put in place by IDCOL will require appropriate regulation and streamlining, including the promotion of technical, managerial, financial, and operational efficiencies. The ultimate aim is to ensure that households receive affordable quality products and services, while the POs can maintain a reasonable profit to sustain their market operation.

Structure of This Book

The book has eight chapters. Chapter 2 describes the current status of Bangladesh's SHS expansion program, including salient features of system operation, as well as program delivery and financing. Chapter 3 reviews the role of electrification in rural development and international experience in using SHS as a complementary solution in remote off-grid areas. Based on the survey data findings, chapter 4 identifies the major drivers of SHS adoption and system capacity selection at the household and village level, while chapter 5 discusses and estimates the welfare benefits. Chapter 6 focuses on SHS market analysis and role of the subsidy, including consumers' willingness to pay and the potential impact of subsidy phase-out. Chapter 7 turns to the quality of PO service and other supply-side issues, along with market constraints to meet future demand. Finally, chapter 8 offers policy perspectives and a way forward.

Notes

1. Details are available at www.worldbank.org/en/news/feature/2014/01/15/lighting-up -rural-communities-in-bangladesh.

2. Investigation of these issues must keep in mind the seasonal effects of longer (summer) and shorter (winter) days.

References

Asaduzzaman, M., Muhammad Yunus, A. K. Enamul Haque, A. K. M. Abdul Malek Azad, Sharmind Neelormi, and Md. Amir Hossain. 2013. *Power from the Sun: An Evaluation of Institutional Effectiveness and Impact of Solar Home Systems in Bangladesh*. Report submitted to the World Bank, Washington, DC.

Barnes, Douglas F., ed. 2007. *The Challenge of Rural Electrification: Strategies for Developing Countries*. Washington, DC: Resources for the Future Press.

BBS (Bangladesh Bureau of Statistics). 2011. *Report of the Household Income and Expenditure Survey 2010*. Dhaka: BBS.

BIDS/World Bank. 2012. "Household Survey Data on Impact Evaluation of Solar Home Systems in Bangladesh." Bangladesh Institute of Development Studies and World Bank, Dhaka.

Brass, Jennifer N., Sanya Carley, Lauren M. MacLean, and Elizabeth Baldwin. 2012. "Power for Development: A Review of Distributed Generation Projects in the Developing World." *Annual Review of Environment and Resources* 37: 107–36.

Cabraal, R. Anil, Douglas F. Barnes, and Sachin G. Agarwal. 2005. "Productive Uses of Energy for Rural Development." *Annual Review of Environment and Resources* 30: 117–44.

Dinkelman, Taryn. 2011. "The Effects of Rural Electrification on Employment: New Evidence from South Africa." *American Economic Review* 101 (7): 3078–108.

Friebe, Christian A., Paschen von Flotow, and Florian A. Täube. 2013. "Exploring the Link between Products and Services in Low-Income Markets: Evidence from Solar Home Systems." *Energy Policy* 52: 760–9.

Islam, A. K. M. Sadrul, Mazharul Islam, and Tazmilur Rahman. 2006. "Effective Renewable Energy Activities in Bangladesh." *Renewable Energy* 31 (6): 677–88.

Jacobson, Arne. 2007. "Connective Power: Solar Electrification and Social Changes in Kenya." *World Development* 35 (1): 144–62.

Khandker, Shahidur R., Douglas F. Barnes, and Hussain A. Samad. 2012. "The Welfare Impacts of Rural Electrification in Bangladesh." *Energy Journal* 33 (1): 187–206.

Komatsu, Satoru, Shinji Kaneko, and Partha Pratim Ghosh. 2011. "Are Micro-Benefits Negligible? The Implications of the Rapid Expansion of Solar Home Systems (SHS) in Rural Bangladesh for Sustainable Development." *Energy Policy* 39 (7): 4022–31.

Lay, Jann, Janosch Ondraczek, and Jana Stoever. 2013. "Renewables in the Energy Transition: Evidence on Solar Home Systems and Lighting Fuel Choice in Kenya." *Energy Economics* 40 (C): 350–59.

Nieuwenhout, F. D. J., A. van Dijk, P. E. Lasschuit, G. van Roekel, V. A. P. van Dijk, D. Hirsch, H. Arriaza, M. Hankins, B. D. Sharma, and H. Wade. 2001. "Experience with Solar Home Systems in Developing Countries: A Review." *Progress in Photovoltaics: Research and Applications* 9 (6): 455–74.

Rebane, Kaja L., and Bradford L. Barham 2011. "Knowledge and Adoption of Solar Home Systems in Rural Nicaragua." *Energy Policy* 39 (6): 3064–75.

Siegel, J. R., and Atiq Rahman. 2011. *The Diffusion of Off-Grid Solar Photovoltaic Technology in Rural Bangladesh*. Medford, MA: Center for International Environment and Resource Policy, The Fletcher School, Tufts University.

Wamukonya, Njeri. 2007. "Solar Home System Electrification as a Viable Technology Option for Africa's Development." *Energy Policy* 35 (1): 6–14.

World Bank. 2013. *Global Tracking Framework for Sustainable Energy for All*. Washington, DC: World Bank.

Zerriffi, Hisham. 2011. *Rural Electrification: Strategies for Distributed Generation*. New York: Springer.

CHAPTER 2

Surge in Off-Grid Solar-Powered Homes

Bangladesh has the world's fastest growing, off-grid solar home system (SHS) coverage. The SHS program started in 2003 with a five-year target of 50,000 units, but within a few years it was installing more than 50,000 units per month. In 2012–13 alone, more than 750,000 systems were installed. By the end of 2013, the total number of installed SHS had surged to more than 2.7 million (figure 2.1). And by early 2014, total installations had reached 3 million. Such phenomenal growth has resulted, in part, from the Government of Bangladesh's program intervention implemented by the Infrastructure Development Company Limited (IDCOL) and its partner organizations (POs), with funds provided by the World Bank and other development partners (box 2.1).

Benefits of SHS Installation

Installing a SHS on the rooftop of a house can have immediate impacts: it enables the household to have light after nightfall, makes study easier in the evenings, allows people to watch TV and be informed of many useful and socially desirable things happening around them, and perhaps be inspired to take part in such activities (box 2.2). Furthermore, it can lower levels of household air pollution (HAP) through reduced use of kerosene, and may even generate extra income by renting charger services for mobile phones. Solar electricity also has the potential positive externality of substituting for fossil fuels in electricity generation and thus contributing to lowering carbon dioxide (CO_2) emissions and the harmful effects of climate change.

Organization of Program Institutions

The institutional organization of Bangladesh's SHS program comprises a well-structured network of partners whose well-defined roles and responsibilities ensure the flow of funding, technical standards for products, quality of

Figure 2.1 Accelerated Growth in Bangladesh's SHS Installations

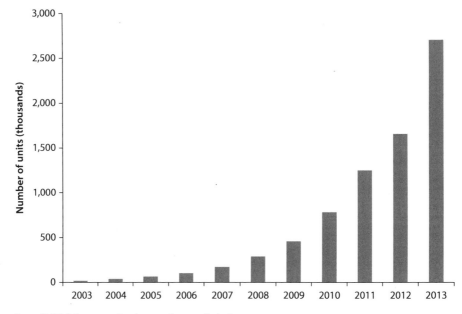

Source: IDCOL (Infrastructure Development Company Limited).
Note: Figures are cumulative. SHS = solar home system.

Box 2.1 Accelerating Energy Access in Bangladesh: RERED II

The second phase of the IDA-funded Rural Electrification and Renewable Energy Development Project (RERED II) is supporting the world's fasting growing solar home system (SHS) program. By 2011, the Infrastructure Development Company Limited (IDCOL), the project's implementing agency, had overseen the installation of 1 million systems. By early 2014, that number had tripled, benefiting nearly 15 million rural people or 10 percent of the country's population. IDCOL has set a target of installing another 3 million systems within two years.

RERED II builds on the success of RERED I, initiated in 2003, which started support to the SHS program in Bangladesh. In addition to IDA funding, the SHS program has been supported by the World Bank's Global Environment Facility (GEF), and is currently funded by the Global Partnership on Output-Based Aid (GPOBA), U.S. Agency for International Development (USAID), German Society for International Cooperation (GIZ), Reconstruction Credit Institute (KfW), Asian Development Bank (ADB), Islamic Development Bank (IDB), Japan International Cooperation Agency (JICA), and Department for International Development (DFID).

Sources: IDCOL; http://www.worldbank.org/en/news/feature/2014/01/15/lighting-up-rural-communities-in-bangladesh.

Box 2.2 How Does a Solar Home System Work?

A solar home system (SHS) offers households in such developing countries as Bangladesh a convenient supply of electricity for lighting and running small appliances (e.g., small television set, radio, and mobile phone charger) for about 3–5 hours a day, using energy from sunlight. Typically, an SHS consists of a small solar photovoltaic (PV) panel, charge controller, battery, compact fluorescent lamp (CFL) or light-emitting diode (LED) lights, and a universal outlet for charging cell phones and small appliances (figure B2.2.1).

Figure B2.2.1 Solar Home System in Action

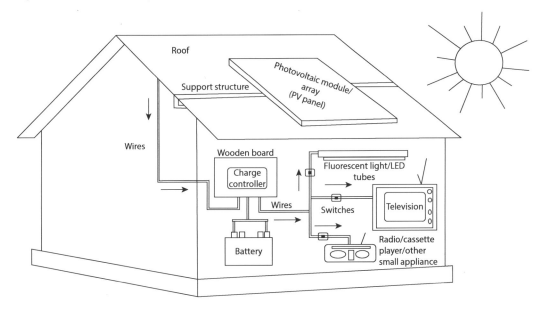

The **solar panel**, also called the **photovoltaic (PV) module**, is the heart of any SHS. Usually installed on the roof of a house at an angle designed to collect maximum sunlight, it converts sunlight into electrical energy. The rechargeable **battery** stores electricity for use at night and on cloudy days and provides the voltage needed to run appliances; in Bangladesh, appliances are designed for 12 volt (V) operation. The **charge controller**, positioned between the solar panel and the battery, protects the battery against overcharging (e.g., on bright sunny days) and discharging below a certain cut-off voltage, which can cause permanent damage. **Watt-peak (Wp)** is the unit of measure used to express the capacity or power generated by the SHS. The capacity range for most SHS units installed in Bangladesh is 20–120 Wp. A system with a 50 Wp capacity can power four lights, a mobile phone charger, and a television set.

Source: IDCOL.

Figure 2.2 Structure of IDCOL-Administered SHS Program

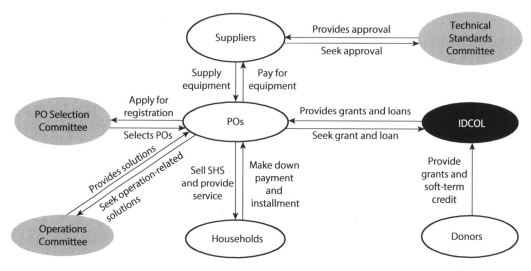

Source: IDCOL.
Note: PO = partner organization; SHS = solar home system.

installations, and after-sales support (figure 2.2). IDCOL recruits the POs, who are responsible for selecting potential SHS buyers in off-grid areas, installing the systems, providing after-sales service and maintenance, and developing a robust market chain. The PO selection committee of IDCOL screens the POs against clear eligibility criteria for inclusion in the IDCOL program. Prior to installation, the technical standards committee approves the suppliers and SHS equipment used. Once the systems are installed, IDCOL conducts a physical verification. It also conducts a technical audit to ensure that only equipment approved by the technical standards committee has been used. The operations committee is responsible for program oversight and providing the POs operational solutions. Donors cover the cost of program administration.

In addition to financing credit operations, IDCOL conducts training and awareness-raising activities for its staff, the POs, and customers. Training covers SHS installation, maintenance and troubleshooting, and market development. Awareness-raising activities include development and distribution of publicity materials to popularize SHS use among rural households throughout the country.

Delivery and Financing Scheme

IDCOL has developed an innovative, partially subsidized SHS delivery and financing scheme, which has proven quite effective in reaching its clientele base. To keep system prices affordable and ensure sustainability beyond the program intervention, IDCOL provides the POs with capital buy-down grants; through market competition, the grants are passed on to household buyers in the form of a lower unit price. Buyers also are offered microcredit financing to make

SHS affordable. These incentives work together to create a robust and regulated rural market chain that ensures quality products that meet safety standards and repair and maintenance facilities with locally available spare parts.

Incentives for Partner Organizations

IDCOL provides the POs several direct incentives that encourage them to lower the unit price to the extent possible. Two types of grants are provided[1]: (a) buy-down grants to reduce household-level costs and promote systems in remote areas and (b) institutional development grants to build capacity of the smaller POs. In addition, the POs can refinance the credit extended to households; they receive a soft loan at a flat interest rate of 6–9 percent for six-to-eight years for 70–80 percent of the credit extended to customers, thus benefiting from the lower interest rate and longer repayment period (table 2.1). Once the POs receive the IDCOL grants and credit refinancing, they pay the suppliers.

Microcredit Financing for SHS Buyers

A vital factor contributing to the success of IDCOL's operation is its microcredit financing mechanism (Siegel and Rahman 2011). For example, a 50 Wp SHS typically costs a household about US$375, a hefty sum in rural Bangladesh (Chakrabarty and Islam 2011; Komatsu, Kaneko, and Ghosh 2011). To make the system affordable, IDCOL requires households to make a 10 percent down payment to the POs and spread installment payments, at a flat 12 percent interest rate, over three years (table 2.1). Once the down payment is received, the POs enter into a sales/lease agreement for microcredit lending with the buyers, the provisions of which are approved by IDCOL. The POs also make a sales agreement with suppliers to get the SHS units and necessary parts and accessories on credit.

Issues for Further Analysis

In this context, a variety of major issues need to be examined. For example, what are the differential costs borne by the clients? What is the distribution of grant benefits between poor and non-poor households? How sustainable is the rural SHS market beyond the project life? What is the level of household customer

Table 2.1 Household Financing Mechanism for 50 Wp SHS

Item	Cost (US$)	Three-year financing terms (US$)	
Market price	400.00	Loan amount	337.50
Buy-down grant to PO (subsidy)	25.00		
Price to household	375.00	Total interest at 12% annually (flat)	
Household down payment (10%)	37.50	for three years	121.50
Household loan extended by PO	337.50	Total household payment	459.00
IDCOL refinancing to PO (80%) at 6% annual rate (flat) for six years	270.00	Monthly household installment	12.75

Source: IDCOL.
Note: PO = partner organization; Wp = watt-peak.

satisfaction with the quality of products and services offered? Do the benefits realized differ among the POs, and do the accrued benefits to households vary because of PO characteristics (e.g., size or concentration of operation in an area)? What measures are being taken by the POs to mitigate environmental damage resulting from the disposal of compact fluorescent lamp (CFL) bulbs, batteries, and charge controllers?

The capital buy-down grants are gradually being reduced, which will induce the POs to change their operating systems to cut costs, or alternatively increase costs to customers. To test the impact on clients, information must be collected and analyzed on PO characteristics, operational costs, staffing quality, and supervisory capacity.

Distribution of PO Market Share

Since the IDCOL-administered SHS program began in 2003, the number of registered POs operating in rural Bangladesh has risen to about 49. Many POs are microcredit lenders, which makes program delivery quite efficient. Market share is dominated by Grameen Shakti; as of March 2014, it accounted for 60 percent of all installations. This is not surprising since Grameen Shakti is part of the Grameen family of organizations, including Grameen Bank, which has the country's most extensive microcredit network and resource base. In March 2014, Rural Services Foundation (RSF), the second largest PO, accounted for 20 percent of installations (figure 2.3).

Figure 2.3 Distribution of SHS Installations, by Partner Organization

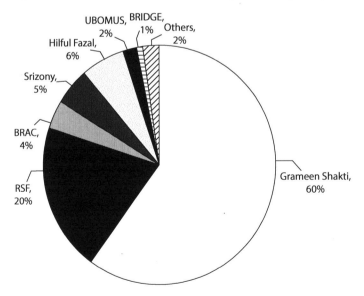

Source: IDCOL.
Note: Figures are cumulative as of March 2014. IDCOL = Infrastructure Development Company Limited; RSF = Rural Services Foundation; SHS = solar home system; UBOMUS = Upokulio Biddutayan O Mohila Unnayan Samity.

Technical Quality of Installations and Monitoring

IDCOL has played a pivotal role in promoting SHS in rural Bangladesh through the POs and has maintained a technical committee to ensure adherence to standards. Despite the absence of a formal regulatory body, IDCOL's approach has ensured better-quality units compared to other market-based suppliers. Thus, one would expect a standardized technical quality of installations.

Problems with technical standards have been noted in the literature. Chowdhury et al. (2011), for example, finds both technical and organizational shortcomings. Design problems include wrongly-sized system components that underperform against approved guidelines and specifications; in some cases, suboptimal placement of key components pose health and safety risks. Guidelines and specifications need more frequent updates to keep pace with technology advances and new research findings, while field offices require appropriate testing facilities and technical training of staff. Additional concerns include the decline in component quality of recent installations and pollution linked to battery disposal (Gottesfeld and Cherry 2011).

To better understand how these issues influence SHS performance and impact client households, this study assessed the technical quality of the installed systems and their operation. Information was gathered from both the POs (headquarters and branch-office levels) and households.

Market Size and Potential Demand

The total potential market for SHS in rural Bangladesh involves various factors related to system finance, subsidy, costs to clients and the POs, and quality of service. Assessing market potential must account not only for technical potential; it must also consider practical intervening factors (e.g., income growth of households, higher consumer demand for energy over time, and expansion of rural grid electrification). Information from the Bangladesh Rural Electrification Board (REB) on national plans for grid extension in rural areas can help to determine future SHS potential in off-grid areas. Results from the large household survey can help to estimate future demand. The market study findings may indicate how, by building on existing markets or creating responsive market structures supported by appropriate financial services, sustainable large-scale SHS dissemination can contribute significantly to rural electrification.

Various supply-and-demand constraints make it unlikely that the total potential market will be targeted. The POs, for example, may consider the current off-grid areas being served as saturated and look elsewhere for SHS opportunities. Determining the potential market on which the POs can base their future expansion plans requires considering the levels of subsidy that may no longer be offered as their SHS service matures. The uncertain quality of grid-electrification service suggests the likelihood of future business demand from grid-connected households that can afford non-subsidized SHS as a backup supply.

Carbon Emissions Reduction

As a least-developed country, Bangladesh is exempt under the rules of the United Nations Framework Convention on Climate Change from any greenhouse gas (GHG) mitigation activity. However, through its submission to the Bali Action Plan at the 13th Conference of the Parties (COP) in 2007, it is committed to keeping GHG emissions as low as possible, provided that adequate financial and technological help are available. Decisions made later at the 2011 Durbin COP were for emissions-reduction actions to be taken by all countries by 2020. By avoiding fossil-fuel electricity generation, Bangladesh's encouraging SHS program fits well with both its earlier commitments and later international developments.

However, there are two caveats. First, it may be argued that avoided carbon must be compared on a life-cycle basis, rather than simply in terms of direct avoidance of the carbon released (Fthenakis and Kim 2011). Second, changing patterns of energy consumption may alter the totality of the country's emissions, but the direction of consumption is unclear. That said, this study estimates direct avoided emissions using data on energy consumption collected at the household level as a first step toward understanding the potential of the SHS program for carbon emissions reduction.

Concluding Remarks

Bangladesh now faces the challenge of providing its citizens sufficient electricity, and demand is increasing. Before deciding on further expansion of the SHS option in off-grid rural areas, policy makers need a better grasp the technology's effectiveness in serving people's direct electricity needs and improving their quality of life. They also need to know whether the observed welfare impacts and future market potential warrant continued subsidy support. Before tackling these issues, the next chapter reviews international lessons in utilizing off-grid SHS to complement grid-based electrification.

Note

1. For each unit installed, IDCOL gives the POs two types of grants: (a) a capital buy-down grant (subsidy) of 10 percent on average to lower unit costs and (b) an institutional development grant for building institutional capacity; however, in cases where units are sold for cash, the subsidy is not provided since the purpose is to create and promote a robust, rural SHS market. Over time, both types of grants have declined; for example, in 2003–04, the total grant per unit was US$90 ($70 capital buy-down and $20 institutional development); by 2013, the total grant had dropped to $23 ($20 buy-down for only smaller system and $3 institutional development) grant only for new POs.

References

Chakrabarty, Sayan, and Tawhidul Islam. 2011. "Financial Viability and Eco-Efficiency of the Solar Home Systems (SHS) in Bangladesh." *Energy* 36 (8): 4821–27.

Chowdhury, Shahriar A., Monjur Mourshed, S. M. Raiyan Kabir, Moududul Islam, Tanvir Morshed, M. Rezwan Khan, and Mohammad N. Patway. 2011. "Technical Appraisal of Solar Home Systems in Bangladesh: A Field Investigation." *Renewable Energy* 36 (2): 772–78.

Fthenakis, V. M., and H. C. Kim. 2011. "Photovoltaics: Life-Cycle Analyses." *Solar Energy* 85 (8): 1609–28.

Gottesfeld, Perry, and Christopher R. Cherry. 2011. "Lead Emissions from Solar Photovoltaic Energy Systems in China and India." *Energy Policy* 39 (9): 4939–46.

Komatsu, Satoru, Shinji Kaneko, and Partha Pratim Ghosh. 2011. "Are Micro-Benefits Negligible? The Implications of the Rapid Expansion of Solar Home Systems (SHS) in Rural Bangladesh for Sustainable Development." *Energy Policy* 39 (7): 4022–31.

Siegel, J. R., and Atiq Rahman. 2011. *The Diffusion of Off-Grid Solar Photovoltaic Technology in Rural Bangladesh*. Medford, MA: Center for International Environment and Resource Policy, The Fletcher School, Tufts University.

CHAPTER 3

Solar Energy's Role in Rural Electrification: International Experience

Electrification is a powerful instrument for raising rural productivity and improving rural households' quality of life (Cabraal, Barnes, and Agarwal 2005; Dinkelman 2011; Khandker, Barnes, and Samad 2012). Along with access to rural markets and credit, supportive government policies, and other complementary conditions, electricity is an essential input in the development of small rural industries (Barnes 2014). In terms of social effects, women and children benefit the most. Rural households that obtain an electricity connection use it initially for lighting, which allows for evening reading and study, especially for school-going children. Even though newly electrified households are likely to continue using traditional biomass cookstoves, women household members often spend less time collecting fuelwood and preparing meals, achieving a better balance of household chores, paid work, and leisure (Barnes 2007).

Lack of access to electricity is considered a major impediment to development; yet the challenge of rural electrification is daunting for many countries. Large investment capital is required; investment in transmission lines is expensive, particularly when targeted populations are located in remote areas with difficult terrain. Distribution companies often have a disincentive to serve sparsely populated rural areas, where prices are set low and poorer households exhibit low levels of electricity demand. Politicians may interfere with the orderly planning and running of programs, insisting on connecting favored constituents first and preventing the disconnection of customers not paying their bills. Poorly designed subsidies can lead distribution companies away from focusing on quality of customer service. In addition, individual farmers may cause difficulties over rights-of-way for the construction and maintenance of electricity lines (Barnes 2007).

To achieve universal electrification, many governments have turned to off-grid solutions utilizing photovoltaic (PV)-based, solar home systems (SHSs) to complement grid-based electrification, particularly in geographically inaccessible areas (Brass et al. 2012; Jacobson 2007; Wamukonya 2007; Zerriffi 2011). In sparsely populated areas with low electricity loads, SHS is often considered the least expensive electrification option. By 2020, it is believed that developing nations will comprise the world's largest solar markets. By harnessing improving levels of efficiency in solar technology, those countries abundant in solar exposure have the potential to create a bright renewable-energy future free from much of the toxic consequences of fossil-fuel generation (Freling and Ramsour 2010). With sustained and aggressive public support, SHS could provide cost-effective basic electricity to a substantial share of rural households while improving both the local and global environment (Duke and Kammen 2003).

Development Benefits to Rural People

Small-scale distributed SHS offers off-grid households in poor rural areas an array of socioeconomic benefits. Recent decentralized SHS initiatives in Sub-Saharan Africa, which have replaced kerosene lighting, have significantly improved the health and educational outcomes of rural households and reduced lighting expenses, leading to higher levels of disposable income (Schultz and Doluweera 2011). The availability of high-quality lighting during evening hours makes it possible to cook after dark, freeing up women's time for income-generating activities earlier in the day. The indirect influence on income generation has been more notable in West Africa than in southern Africa (Nieuwenhout et al. 2001). In remote villages of northern Benin, where national grid extension is not economically viable, solar-powered drip irrigation projects implemented by the Solar Electric Light Fund (SELF) have overcome local residents' concern over food security during the dry season, helping them break out of the cycle of poverty. Additional off-grid applications have included pumps to provide fresh drinking water and power for schools and health clinics, street lighting, community centers, and WiFi networks (Freling and Ramsour 2010).

Overcoming Financial Hurdles

Lack of financial services to cover SHS purchase is a serious bottleneck to adoption for poorer rural households, who must allocate their limited resources to other essential livelihood items, as well as electricity (Friebe, von Flotow, and Täube 2013; Nieuwenhout et al. 2001). Many banks are either unwilling to lend to the poor, perceiving them as high-risk, or charge exorbitant interest rates with a large down payment (Dahlke 2011). Projects that utilize leasing or hire-purchase arrangements have had higher sustainability rates than donor-supported and heavily subsidized projects and have achieved greater market penetration than cash-sale approaches. Under such credit arrangements, household consumers make an initial down payment based on affordability and pay the balance of capital costs in installments over a specified period.

In South Asia, Grameen Shakti, a nonprofit subsidiary of Grameen Bank in Bangladesh, sold about 42,000 systems in 2005 using a 15 percent down payment with a three-year credit period, achieving a 90 percent recovery rate (Anisuzzaman and Urmee 2006) (box 3.1). In India, SELCO, a for-profit social enterprise based in Karnataka, has succeeded in selling, servicing, and financing more than 115,000 SHS since 1995. Using a microcredit approach, SELCO has forged partnerships with rural and commercial banks, nongovernmental organizations (NGOs), and farmer cooperatives. Products are customized, with an emphasis on quality installation, user education, and after-sales service.

In Central America, a SHS pilot project in Costa Rica supported by Light Up the World (LUTW), a Canadian nonprofit organization, succeeded by basing loan repayment on existing expenditure patterns; that is, households diverted the amount of money they spent each month on kerosene and candles to repay the cost of the SHS (Schultz and Doluweera 2011). Various for-profit enterprises have also succeeded in extending SHS to off-grid rural areas. Tecnosol, based in Nicaragua, with operations in El Salvador and Panama, has installed more than 40,000 SHS units in Central America using a microcredit approach. In the absence of microfinance availability, Soluz, an enterprise based in the Dominican Republic, has pioneered a financial approach based on non-subsidized micro-rentals. Quetsol, based in Guatemala, has provided rural customers simple SHS units using a microcredit approach that includes partnerships with banks, microfinance institutions, and local NGOs (Dahlke 2011).

Box 3.1 Grameen Shakti: A Formidable Partner Organization

Grameen Shakti (meaning "village energy or power" in Bengali) is a not-for-profit company in Bangladesh established in 1996 under the Grameen Bank. It aims to promote and supply renewable energy products in rural Bangladesh that are affordable to poor households. Benefiting from the Grameen Bank's 40 years of experience in microcredit lending and extensive countrywide network, Grameen Shakti has become the largest partner organization (PO) of the Infrastructure Development Company Limited (IDCOL). By late 2012, it had installed more than a million solar home systems (SHS) in rural Bangladesh and now accounts for well over half of all IDCOL-supported installations. For its work in renewable-energy technology, Grameen Shakti received the European Solar Award in 2003 and the United Kingdom's Ashden Award for Sustainable Energy in 2006. It also works to promote biogas plants and improved cookstoves.

Grameen Shakti's SHS-promotion work extends beyond installations. It also is dedicated to local capacity building and employment generation, particularly for women. By late 2013, it had established 46 technology centers throughout Bangladesh where customers, mainly women, are trained as technicians to service and repair SHS in their local areas (see photos B3.1.1 and B3.1.2). To date, more than half a million customers have been trained, facilitating the generation of large-scale rural employment.

box continues next page

Box 3.1 Grameen Shakti: A Formidable Partner Organization *(continued)*

Photo B3.1.1 Women in Training at Grameen Technology Centre

Photo B3.1.2 Woman Working for Grameen Technology Centre

Source: Grameen Shakti.
Photos: Sarah Butler-Sloss/Ashden. © Ashden. Used with permission. Further permission required for reuse.

Technical Considerations

Project success requires a well-designed SHS (box 2.2), using quality component products, along with technically sound installation and after-sales service. Batteries and lights cause most technical problems, and proper functioning of charge controllers is essential to ensuring a high battery lifetime. Battery lifetimes vary significantly by project. Those designed for solar application may last up to five years, while locally produced automotive batteries typically last only one-to-three years. Lighting quality may vary considerably between comparable lamp types. Breakdowns of compact fluorescent lamps (CFLs) and other inexpensive components may create dissatisfaction among users and reduce their willingness to continue making payments. Design improvements should be based on field experience, including household users' feedback (Nieuwenhout et al. 2001).

Building a Thriving Off-Grid Market

In addition to the need for technologically suitable products, robust market development requires making systems compatible with the social, cultural, and economic activities of targeted households. A one-size-fits-all approach is unlikely to succeed. Rather, SHS businesses should offer a suite of products that fit consumers' diverse needs and choices, including smaller module capacities (i.e., 10–30 watt-peak [Wp]). Partner organizations (POs) should be carefully selected with clearly defined roles and responsibilities, and delivery and financing schemes should be designed to ensure the viability of financing institutions and intermediaries.

Other ingredients vital to long-term success include establishing supply networks for products and services, including a functioning maintenance and service scheme; launching campaigns to build household awareness; and training customers in SHS operation and basic troubleshooting (Schultz and Doluweera 2011). India's Barefoot College, which has trained thousands of women as solar engineers, has benefited some 190,000 people in 16 states across India and has been replicated in numerous countries throughout South Asia, Africa, and Latin America. In Bangladesh, Grameen Shakti has started a network of technology centers, managed mainly by women engineers who train other women as solar technicians (box 3.1). Investing in community outreach and demonstrations is especially important in villages with no prior knowledge of SHS where politicians have made unmet promises to extend the national grid (Dahlke 2011). Market transparency can be enhanced by empowering household members via independent consumer organizations that compare various commercially available products (Nieuwenhout et al. 2001).

Beyond Off-Grid Markets

Traditionally, energy planners have viewed solar energy as having enormous technical potential and offering high economic returns in off-grid areas but economically unattractive as a contributor to large-scale power generation.

But recent improvements in the global cost of solar energy technology suggest that solar PV may already be an attractive alternative to the most expensive conventional generation technologies (e.g., emergency power plants running on heavy fuel oil). A recent case study in Kenya suggests power-sector expansion would be cheaper if utility-scale solar plants, versus additional fossil-fuel plants, were connected to the country's power grid. Bangladesh also has potential for grid-supplied solar electricity (Hossain and Sadrul Islam 2011). In countries with enabling government policies, excellent market penetration for grid-connected solar applications is being achieved (Holm 2013).

Potential for Bangladesh

Providing electricity is central to Bangladesh achieving its vision of middle-income status by 2021, yet nearly three-fifths of rural households still lack access to electric power. The country's limited ability to generate and distribute enough grid-based electricity to meet growing demand (Barnes 2007; Zerriffi 2011), combined with its ample sunshine and high levels of energy poverty, suggests a large potential market for SHS in poorer off-grid areas. Chapters 6 and 7 discuss the financial, as well as technical and institutional, hurdles that will need to be overcome along the way. Before turning to these issues, chapter 4 identifies the factors that influence SHS adoption, while chapter 5 measures the welfare gains to households.

References

Anisuzzaman, M., and Tania P. Urmee. 2006. "Financing Mechanisms of Solar Home Systems for Rural Electrification in Developing Countries." http://www.researchgate.net.

Barnes, Douglas F., ed. 2007. *The Challenge of Rural Electrification: Strategies for Developing Countries.* Washington, DC: Resources for the Future Press.

Barnes, Douglas F. 2014. *Electric Power for Rural Growth: How Electricity Affects Life in Developing Countries.* 2nd ed. Washington, DC: Energy for Development. http://www.energyfordevelopment.com.

Brass, Jennifer N., Sanya Carley, Lauren M. MacLean, and Elizabeth Baldwin. 2012. "Power for Development: A Review of Distributed Generation Projects in the Developing World." *Annual Review of Environment and Resources* 37: 107–36.

Cabraal, R. Anil, Douglas F. Barnes, and Sachin G. Agarwal. 2005. "Productive Uses of Energy for Rural Development." *Annual Review of Environment and Resources* 30: 117–44.

Dahlke, Steve. 2011. "Solar Home Systems for Rural Electrification in Developing Countries: An Industry Analysis and Social Venture Plan." ENTR 311 Social Entrepreneurship, Terri Barreiro College of St. Benedict and St. John's University. http://www.csbsju.edu.

Dinkelman, Taryn. 2011. "The Effects of Rural Electrification on Employment: New Evidence from South Africa." *American Economic Review* 101 (7): 3078–108.

Duke, Richard D., and Daniel M. Kammen. 2003. "Energy for Development: Solar Home Systems in Africa and Global Carbon Emissions." In *Climate Change for Africa: Science, Technology, Policy and Capacity Building*, edited by Pak Sum Low, 250–66. Dordrecht, the Netherlands: Kluwer Academic Publishers.

Freling, Robert A., and David Lawrence Ramsour. 2010. "Shining Light on Renewable Energy in Developing Countries." http://www.definitivesolar.com.

Friebe, Christian A., Paschen von Flotow, and Florian A. Täube. 2013. "Exploring the Link between Products and Services in Low-Income Markets: Evidence from Solar Home Systems." *Energy Policy* 52: 760–9.

Holm, Dieter. 2013. "Renewable Energy Future for the Developing World." In *Transition to Renewable Energy Systems*, edited by D. Stolten and V. Scherer. Weinheim, Germany: Wiley-VCH Verlag GmbH & Co. KGaA.

Hossain, Md. Alam, and A. K. M. Sadrul Islam. 2011. "Potential and Viability of Grid-Connected Solar PV System in Bangladesh." *Renewable Energy* 36 (6): 1869–74.

Jacobson, Arne. 2007. "Connective Power: Solar Electrification and Social Changes in Kenya." *World Development* 35 (1): 144–62.

Khandker, Shahidur R., Douglas F. Barnes, and Hussain A. Samad. 2012. "The Welfare Impacts of Rural Electrification in Bangladesh." *Energy Journal* 33 (1): 187–206.

Nieuwenhout, F. D. J., A. van Dijk, P. E. Lasschuit, G. van Roekel, V. A. P. van Dijk, D. Hirsch, H. Arriaza, M. Hankins, B. D. Sharma, and H. Wade. 2001. "Experience with Solar Home Systems in Developing Countries: A Review." *Progress in Photovoltaics: Research and Applications* 9 (6): 455–74.

Schultz, C., and G. Doluweera. 2011. "Best Practices for Developing a Solar Home Lighting System Market." *Journal of African Business* 12 (3): 330–46.

Wamukonya, Njeri. 2007. "Solar Home System Electrification as a Viable Technology Option for Africa's Development." *Energy Policy* 35 (1): 6–14.

Zerriffi, Hisham. 2011. *Rural Electrification: Strategies for Distributed Generation*. New York: Springer.

CHAPTER 4

Patterns of SHS Growth and Usage: Survey Data Findings

Assessing the off-grid market potential of a solar home system (SHS) in rural Bangladesh requires a better understanding of the social, economic, and other factors that together influence household adoption. To grapple with these issues, this study collected both primary and secondary data on SHS adoption in off-grid rural areas. The primary data were gathered from three field surveys conducted in mid-2012 by the Bangladesh Institute of Development Studies (BIDS) under the World Bank–supported Rural Electrification and Renewable Energy Development Project (RERED II) (box 2.1). The secondary data were provided by the Infrastructure Development Company Limited (IDCOL), the project's implementing agency.

Overview of Survey Design

In May–June 2012, BIDS conducted a large household survey of SHS adopters and non-adopters across a wide range of household characteristics and welfare information. A total of 4,000 households were surveyed (1,600 treatment and 2,400 control). IDCOL's database, which maintains detailed information on SHS household adoption nationwide, was used for selecting the sampling framework for the treatment households (Asaduzzaman et al. 2013).

The survey used a multi-stage sampling technique, with "village" and "household" as the respective primary and secondary sampling units. All seven divisions of the country were covered to ensure geographic representativeness; of these, 16 districts were selected, ordered according to their SHS concentrations, with at least one district from each division included. From each of the selected districts, two subdistricts, locally known as *upazilas*,[1] were chosen, based on their concentrations of SHS installations; thus, a total of 32 upazilas were selected. Similarly, from each upazila, two treatment villages were selected at random (annex 4A). From each village, 25 SHS adopter households were randomly selected, for a total of 1,600 treatment households. In each of these villages, 6–7 households that had not yet adopted SHS were selected, for a total of 400

Table 4.1 Distribution of Survey Sample, by Division

Division	District	Upazila	Village Treatment	Village Control	Household Treatment	Household Control
Dhaka	3	6	12	12	300	450
Chittagong	2	4	8	8	200	300
Khulna	3	6	12	12	300	450
Rajshahi	1	2	4	4	100	150
Rangpur	1	2	4	4	100	150
Barisal	4	8	16	16	400	600
Sylhet	2	4	8	8	200	300
Total	16	32	64	64	1,600	2,400

Source: BIDS/World Bank 2012.

non-adopter, control households in the treatment villages. The remaining 2,000 control households were selected from control villages, for a total of 2,400 control households (table 4.1).

In each upazila where they operate, the partner organizations (POs) maintain a list of SHS adopter households by village. From these lists for the already selected upazilas, two villages with the least concentrations of SHS households were selected as control villages. The two control villages were from the same upazilas as the two treatment households, making it likely that these two sets of villages would have similar observable characteristics. From each of the control villages, 31–32 households without SHS were randomly selected, for a total of 2,000. Since SHS households were overdrawn in the sample, all data analyses were weighted so that the findings would be representative of the country's off-grid rural areas (annex 4A).

The data collected covered household assets, income, expenditure, and housing conditions and sanitation. Data also was collected on the education, employment, and health of household members and time-use patterns of adult household members. Detailed data were captured on households' energy-use patterns, including consumption of various energy sources and expenditure on appliances. SHS owners were questioned about the purchase of their units, usage, system history, and quality of service.

In addition to the household survey, a community survey was conducted in each village and a questionnaire was fielded to all PO branch offices to investigate the cost-effectiveness of the SHS technology and delivery system for household owners. The community survey collected data on village infrastructure, prices of consumer goods, daily wages, and energy prices, while the PO questionnaire assessed the performance of all POs operating in the sample villages.

Growth Trend in SHS Adoption

By late 2012, the SHS adoption rate for off-grid rural households had reached 8 percent on average. The divisions of Barisal and Sylhet exhibited the highest intensities of household use, at 13.4 percent and 13.2 percent, respectively; while

Table 4.2 Extent of SHS Adoption in Rural Bangladesh by Region, 2012

Division	Household adoption rate (%)
Dhaka	7.8
Chittagong	10.2
Khulna	7.6
Rajshahi	3.9
Rangpur	3.3
Barisal	13.4
Sylhet	13.2
All regions	8.0
N	4,000

Source: BIDS/World Bank 2012.

Figure 4.1 SHS Adoption by Year from Sample Data

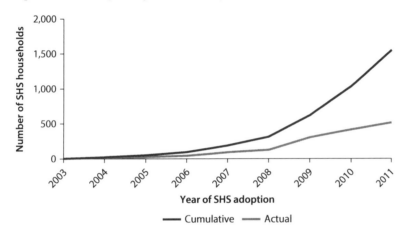

Source: BIDS/World Bank 2012.
Note: SHS = solar home system.

the lowest adoption rates were found in Rajshahi and Rangpur, at 3.9 percent and 3.3 percent, respectively (table 4.2).

The overall trend in cumulative household adoption from the study sample is quite similar to the nationwide adoption rates reported in chapter 2 (figure 2.1). The cumulative figures clearly reflect the surge in SHS adoption in recent years, particularly from 2009 onwards (figure 4.1).

Drivers of SHS Adoption

Few studies have examined the determinants of SHS adoption, though many claim off-grid solutions as a better option than coal-based electrification because it is a renewable energy source. Grid-connected households in more developed countries sometimes adopt SHS, suggesting its use as a backup power source for households willing to pay. Findings from previous studies in Bangladesh and elsewhere suggest that, when SHS is made available in the local market, the early adopters tend to be wealthier households (see, e.g., Siegel and Rahman 2011). This is not surprising, given the high upfront investment and ongoing costs involved. By contrast, poorer households usually hesitate to adopt applications that require high upfront costs, even though they could save money over the long run. An examination of SHS adoption rates by landholding confirms these findings.[2] Only 10 percent of low-to-medium landholders (i.e., 250 decimals or less) adopted SHS, while the rate surged for large landholders (i.e., more than 250 decimals). For households owning more than 500 decimals (about 5 acres), the adoption rate could exceed 25 percent (figure 4.2).

In addition to landholding, a variety of household- and village-level factors could influence a household's decision to adopt a SHS. To capture the effects of such factors, the study estimated SHS demand using the following equation:

$$S_{ij} = \alpha + \beta X_{ij} + \gamma V_j + \varepsilon_{ij}, \tag{4.1}$$

where S_{ij} indicates whether ith household living in jth village has a SHS unit (a binary variable with a value of 1 when a household has a system and 0 otherwise). X_{ij} represents household-level control variables, including measures of household assets and education of household members. V_{ij} represents village-level exogenous variables (e.g., infrastructure and prices, including alternative energy sources). Finally, e_{ij} equals an unobserved random error, while b and g are parameters to be determined. Because of the binary nature of SHS adoption, a probit model is applied to the SHS adoption equation. Table 4.3 presents the results of the SHS demand regression and summary statistics for the explanatory variables used in the estimation. The effects of various factors on SHS adoption from both the regression results and descriptive trends are presented below.

Figure 4.2 SHS Adoption Rates and Household Landholdings

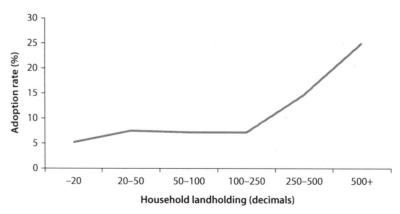

Source: BIDS/World Bank 2012.
Note: SHS = solar home system.

Table 4.3 Regression Estimate of SHS Demand

N = 4,000

Explanatory variable	Probit estimate	Mean and standard deviation
Sex of household head (1 = male, 0 = female)	−0.038** (−3.48)	0.945 (0.229)
Age of household head (years)	−0.001** (−2.60)	46.3 (12.9)
Education of household head (years)	0.006** (5.98)	3.31 (3.90)
Housing structure is mostly or completely brick-built (1 = yes, 0 = no)	0.071** (4.00)	0.035 (0.185)
Housing uses sanitary latrine (1 = yes, 0 = no)	0.015* (1.78)	0.367 (0.482)
Housing uses drinking water from arsenic-free tube well (1 = yes, 0 = no)	0.016* (1.68)	0.401 (0.490)
Log household land asset (decimals)	0.008** (2.45)	128.3 (236.9)
Log household non-land asset (thousand Tk)	0.020** (5.90)	1,973.8 (7,301.3)
Village price of fuelwood (Tk/kg)	0.002 (0.15)	4.15 (1.42)
Village price of kerosene (Tk/liter)	−0.003 (−0.57)	64.9 (1.75)
Log price of SHS (Tk/Wp)[a]	−0.021** (−1.81)	6.91 (0.003)
Village has Grameen Bank microcredit program (1 = yes, 0 = no)	−0.054 (−1.47)	0.712 (0.453)
Village has BRAC microcredit program (1 = yes, 0 = no)	0.075** (2.21)	0.804 (0.397)
Village has other non-microcredit NGO programs (1 = yes, 0 = no)	0.038 (1.16)	0.754 (0.431)
Village is in *char* area (1 = yes, 0 = no)	0.063* (1.96)	0.281 (0.449)
Village is subject to river erosion (1 = yes, 0 = no)	0.090** (2.78)	0.171 (0.376)
Village has paved roads (1 = yes, 0 = no)	−0.014 (−0.57)	0.557 (0.497)
Pseudo R^2	0.220	
Mean and standard deviation of dependent variable	0.080 (0.272)	

Source: BIDS/World Bank 2012.

Note: Regression includes additional control variables, such as village prices of consumer food items and dummy variables for divisional regions. kg = kilograms; NGO = nongovernmental organization; SHS = solar home system; Wp = watt-peak.

a. The SHS price is calculated by dividing the purchase price of the unit by capacity and then taking the village-level average.

Significance level: * = 10 percent, ** = 5 percent.

Household Wealth and Income

We have already seen that households with more land purchase SHS at a higher rate than those with less land. The regression results confirm that landholding has a positive effect on SHS adoption. In addition, SHS owners tend to have higher levels of overall assets, including agricultural machinery and financial resources; a 10 percent increase in non-land assets increases the probability of SHS adoption by 0.2 percentage point. Other influencing factors include the quality of housing structure and household hygienic practices, which reflect income. The study survey data also show that SHS owners earn an average of about US$2,000 per year, which is 80 percent higher than the average for non-owners. Furthermore, SHS adopter households derive a significantly lower percentage of their annual income from agriculture than do non-adopters.

Role of Education and Gendered Dimension

SHS adopter households tend to have better educated members than their non-adopter counterparts. Regression results show that one extra year of education of the household head increases the probability of SHS adoption by 0.6 percentage point. Among SHS owners, more than two-fifths of household heads have completed secondary school or beyond, compared to only one-fifth of non-owner household heads. In adopter households, 76 percent have at least one woman member who has completed primary school, compared to 60 percent for non-adopter households; and 20 percent have women members who have completed secondary school, compared to only 10 percent for non-adopters. Both women and men household heads—62.5 percent and 54.7 percent, respectively—acknowledged the positive role of SHS in facilitating children's education.

Price Effect

The demand for SHS units declines as their price increases, indicating a negative effect of own price. For every Tk 100 increase in unit price (i.e., Tk per watt-peak), the probability of SHS adoption decreases by 2 percentage points.[3] The cross-price effects of SHS substitutes, such as fuelwood and kerosene, are not statistically significant for SHS adoption; that is, their prices do not affect SHS demand.

Other Factors

At the village level, both agroclimatic and socioeconomic features play an important role in SHS adoption. Villages in areas more geographically remote or vulnerable to natural calamities are more likely to exhibit greater household demand. This is not surprising since these are mainly off-grid villages unlikely to receive a grid connection in the foreseeable future—the reason why SHS is heavily promoted in such areas; for example, if a village is located in a *char* area, the probably of SHS adoption increases by 6.3 percentage points.[4] Adopters also tend to live in villages with a strong microcredit presence. In villages where the BRAC microcredit program operates, the probability of SHS adoption increases by 7.5 percentage points; however, the Grameen Bank's presence has no independent effect on SHS adoption.[5]

System Capacity and Appliance Use

It is illuminating to investigate the recent growth trend in SHS adoption for the various capacity systems offered (figure 4.3). The overall trend is similar to the growth rate reported in figure 2.1. The 50 Wp system has seen the fastest growth, followed by the 40 Wp system. Recent price declines have made these capacity systems attractive choices; both allow for using a moderate range of appliances (e.g., lights, TV, and mobile phone charger). Currently, 20 Wp systems are gaining in popularity owing to introduction of a system based on a light-emitting diode (LED) for smaller-capacity units. Despite the enormous growth in SHS adoption, only about 8 percent of off-grid rural households had been reached by late 2011.[6] This means there is ample room for further SHS expansion, and demand is not going to diminish anytime soon.

Predictably, energy consumption from the SHS panel increases with watt-peak size (figure 4.4). With higher-capacity systems, households extend their

Figure 4.3 Trend in SHS Adoption Rate by Capacity, 2004–11

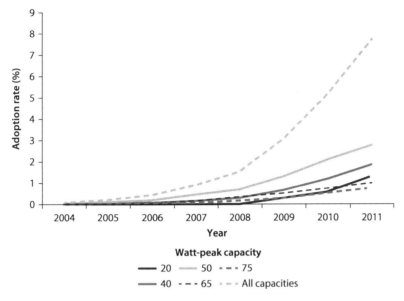

Source: BIDS/World Bank 2012.
Note: Growth in adoption rate is cumulative. SHS = solar home system.

Figure 4.4 Household Energy Consumption, by SHS Capacity

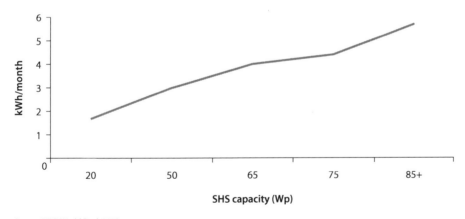

Source: BIDS/World Bank 2012.
Note: kWh = kilowatt hour; SHS = solar home system; Wp = watt-peak.

appliance use beyond lighting, typically purchasing a television set, fan, and other small appliances.

Table 4.4 shows the share of SHS adopters that use various appliances. As expected, there is a positive correlation between watt-peak capacity and the number of lights a SHS supports. For example, a 20 Wp system supports, on average, only one light bulb, whereas a 75 Wp system supports up to five light bulbs. Charger lights, one of the most common appliances, are used by 13 percent of all SHS adopters, while 37 percent use SHS-powered electricity to run a television set.

Table 4.4 Appliance-Use Patterns of Adopter Households, by SHS Capacity

System capacity (Wp)	Share of SHS users (%)	Tube light/CFL (number)	Charger lights/lanterns (%)	TV set (%)
20	17.3	1.0	9.0	5.8
40	23.9	2.4	8.7	33.2
50	36.1	3.3	13.2	46.1
65	12.9	3.8	23.4	50.7
75 and over	9.8	4.7	18.6	49.6
All capacities	100.0	2.9	13.3	37.0

Source: BIDS/World Bank 2012.
Note: CFL = compact fluorescent lamp; SHS = solar home system; Wp = watt-peak.

Table 4.5 Household Energy Consumption and Use, by SHS Adoption

Energy source	SHS adopter households (N = 1,600)	SHS non-adopter households			t-statistics of difference between SHS users and non-users
		SHS villages (N = 400)	Control villages (N = 2,000)	All villages (N = 2,400)	
Consumption (kgOE per month)					
Fuelwood	63.57	55.68	49.61	50.64	4.48**
Non-fuelwood biomass[a]	61.87	61.20	65.48	64.80	−1.27
Kerosene	0.76	2.39	2.33	2.34	−23.62**
	(0.92)[c]	(2.91)[c]	(2.82)[c]	(2.84)[c]	
SHS	0.30	0	0	0	
	(3.56)[d]				
Other sources[b]	0.29	0.02	0.02	0.02	5.36**
All sources	104.32	97.05	97.21	97.18	1.30
Use (% per month)					
Fuelwood	86.5	82.2	79.8	80.2	
Non-fuelwood biomass[a]	78.1	80.0	84.4	83.7	
Kerosene	61.6	97.0	99.1	98.7	
SHS	100.0	0	0	0	
Other sources[b]	52.5	65.0	63.8	64.0	

Source: BIDS/World Bank 2012.
Note: Consumption figures are average values for households that use a particular energy source; households with zero consumption are excluded from the calculation. kgOE = kilograms of oil equivalent; SHS = solar home system.
Significance level: ** = 5 percent.
a. Non-fuelwood biomass includes dung, tree leaves, crop residue, charcoal, jute stick, and briquette.
b. Other sources include liquefied petroleum gas (LPG), candles, dry-cell batteries, storage batteries, and generators; a significant share of households use these energy sources, but only in small amounts.
c. Figures in parentheses show consumption in liters per month.
d. Figure in parentheses shows consumption in kWh per month.

Composition of Energy Consumption and Use

How does SHS adoption change household energy-use patterns in off-grid rural areas? Regardless of whether they adopt SHS, off-grid households depend on biomass and kerosene to meet most of their daily energy requirements (table 4.5).

About four-fifths use fuelwood and other forms of biomass to meet cooking and related needs. SHS adopter households use 13 percent more fuelwood per month than non-adopters (64 versus 51 kgOE), while non-adopters consume 3 percent more non-fuelwood biomass than adopters (65 versus 62 kgOE).[7]

At the same time, the share of kerosene use among SHS adopters is 37 percent less than that of non-adopters (62 versus 99 percent) (table 4.5). Compared to non-adopter households, households that adopt SHS consume an average of 2 liters less kerosene per month (1 versus 3 liters), and this difference is statistically significant. Clearly, SHS adopters do not require kerosene for lighting.[8] In short, SHS adoption changes the composition of household energy consumption, but there is no statistically significant difference in total energy consumption between adopters and non-adopters.

Summing Up

Key socioeconomic characteristics at the household and intra-household level, certain village features, and unit pricing figure prominently in SHS adoption and system capacity selection in rural off-grid Bangladesh. Predictably, wealthier households with higher levels of education adopt higher-capacity units and, as a result, can afford a diverse set of appliances. Poorer households with younger household heads tend to purchase less expensive, smaller-capacity units. While SHS adopters and non-adopters differ little in terms of overall energy consumption, they vary significantly in terms of energy-use composition. Not surprisingly, non-adopters use three times as much kerosene, implying that SHS adoption results in a huge reduction in carbon dioxide (CO_2) emissions when aggregated nationwide. The reduction in CO_2 emissions is quantified in the next chapter, which discusses the various beneficial effects of SHS.

Annex 4A: Household Survey Design

The World Bank–supported household survey was conducted by the BIDS in May–June 2012. A total of 24 field officers and field supervisors were deployed for data collection under supervision of a survey coordinator. This annex describes the sampling framework used in the survey design, including spatial sampling and household selection, as well as the weighting procedure for household adoption rates by division.

Sampling Framework

The survey sample totaled 4,000 households; more control households than treated ones (2,000 versus 1,600) were selected in order to increase the number of matches using propensity-score matching (PSM) and thus increase the robustness of the estimates.

Sample Size Determination

Because households are the main SHS users, the approach used to estimate the sample size was based on confidence level and precision rate. The formula used can be expressed as follows:

$$n_h = [(z^2)*(1-r)*f*k]/[r*p*s*e^2], \tag{4A.1}$$

where n_h is the sample size in terms of number of households to be selected, z is the normal density function that defines the level of confidence desired, r is an estimate of an indicator to be measured by the survey, f is the sample design effect, k is a multiplier to account for the anticipated rate of non-response, p is the proportion of the total population accounted for by the target population, s is the average household size, and e is the relative margin of error.

In this study, the level of significance was set at 10 percent, with the consequent z-value of 1.64. Since SHS penetration was less than 5 percent, the indicator r was set at 0.04. The design effect, denoted by f, was set at 2. The extent of non-response was set at 5 percent. As of the latest national census, the proportion of rural population was about 75 percent; thus 0.75 was used for p. The 2010 Household Income and Expenditure Survey (HIES) indicates a rural household size of 4.53, which was used for s. The extent of relative margin of error was set at 10 percent. Given the values of the parameters, the above formula results in a total sample size of 3,990 households; thus, the actual sample size was determined at 4,000.

Spatial Sampling

To facilitate the comparison between treatment (with SHS) and control (without SHS) households, the data collection was split. Of the sampled 4,000 households, 1,600 had adopted SHS; these treated households were located in villages with relatively high SHS adoption rates. To minimize spillover effects and unobserved heterogeneity vis-à-vis the treated households, data were collected on 2,000 households located in adjacent *union parishads* that had not adopted SHS.[9] Finally, data were collected on another 400 control households located in the same villages as the treated households.

The intervention outcomes of the 1,600 treated households were matched with those of the 2,000 control households from the adjacent villages, which were used to assess impact. By contrast, data collected on the 400 control households from the treated villages were used to assess determinants of program participation and the proximate factors that deter adoption of the technology.

To ensure geographic and PO representativeness, a multi-stage, stratified cluster sampling method was adopted. The sampling frame was based on data provided by the IDCOL on cumulative SHS installations through 2011. In the first stage, 16 districts were selected from the country's 7 administrative divisions (table 4A.1). Selection of districts was based on highest concentrations of SHS installations. However, this selection criterion meant that one division was under-represented—Sylhet had only one district (Sunamganj) in the top

Table 4A.1 Selected Districts, Upazilas, and Treatment Villages

Division	District	Upazila	Treatment villages
Dhaka	Mymensingh	Dhobara	Bakpara, Baligaon
		Haluaghat	Kumarghati, Songra
	Netrokona	Kendua	Jalli, Shivpur
		Kaliajuri	Adampur, Chandpur
	Shariatpur	Bhederganj	Baher Char, Char Bhaga
		Shariatpur Sadar	Kachikata, West Atpara
Chittagong	Chittagong	Banshkhali	Gandamara, Premashia
		Sandwip	Maitbhanga, Rahmatpur
	Noakhali	Hatiya	Dashpara, Sukh Char
		Noakhali Sadar	Kazir Char, West Maiz Chara
Khulna	Bagerhat	Chitalmari	Aruaborni, Barashia
		Mongla	Joykha, Nitakhali
	Khulna	Dumuria	Chitramari, Khorerabad
		Terakhada	Adampur, Kushala
	Satkhira	Shymnagar	Gumantoli, Kamalkati
		Tala	Khanpur, Magura
Rajshahi	Bogra	Sherpur	Khanpur, Simla
		Shivganj	Chakpara, Chandrapukur
Rangpur	Kurigram	Chilmari	Duttar Char, Mudafat Kalikapur
		Rajibpur	Baliamari, Kodalkati
Barisal	Barguna	Amtali	Sonakhali, Tariakata
		Barguna Sadar	Kumrakhali, Sharishamuri
	Barisal	Barisal Sadar	Nalchap, Rajar Char
		Muladi	Alimabad, Charbatamara
	Bhola	Bhola Sadar	Kandokpur, West Char Pata
		Char Fashion	Aminpur, Hazariganj
	Patuakhali	Dashmina	Ali Pura, Bara Gopaldi
		Galachipa	East Neta, Kazi Kanda
Sylhet	Sunamganj	Dharmapasha	Kandapara, Rupnagar
		Jagannathpur	Chilaura, Paragaon
	Hobiganj	Chunarughat	Parkul, Bholarjum
		Lakhai	Begunai, Faridpur
Total	16	32	64

Source: BIDS/World Bank 2012.

16 districts—while another was not represented—Rajshashi had no districts among the top 16. To circumvent this practical problem, two districts from the top 16 list (Chandpur and Tangail) were replaced by Bogra (Rajshahi) and Hobiganj (Sylhet).[10] There is a high variation in SHS concentration across districts. It is highest in Sunamganj, where more than 68,000 systems are installed, and lowest in Bogra, which has only about 21,000 SHS installations. About 58 percent of total installations are concentrated in these 16 districts.

In the second stage, two upazilas from each of the selected districts were selected at random, based on concentration of SHS installations; thus, 32 upazilas

Surge in Solar-Powered Homes • http://dx.doi.org/10.1596/978-1-4648-0374-1

were selected. Similarly, two villages were randomly selected from each upazila, for a total of 64 villages. While upazila selection was proportionate to the size of installations, villages were required to have had SHS installation programs for at least three years and, for ease of sampling, have at least 50 SHS households. When these two criteria were applied, a few of the 32 upazilas had only one village that met these conditions, even after repeated random draws. Several attempts were made, and the best-case scenario left two upazilas—one in Barisal and the other in Kurigram—with one village. These two upazilas were revisited, and the selection criteria were relaxed. From each of these upazilas, the village with the highest SHS concentration, which was less than 50, was selected to compensate for the shortage of villages. From each of the other upazilas, two villages were selected at random.

Household Selection

A stratified random sampling method was applied to choose households from the selected villages. The basis for stratification was the SHS capacity size (10, 20, 40, 80, 130, and 150 Wp) installed in the villages under IDCOL financial and technical support. To ensure representation of variously sized panels, 25 households were proportionally allocated among size groups using the following formula:

$$n_h = \frac{n}{N} N_h, \tag{4A.2}$$

where, n_h represents the number of households with hth size of SHS, n is the total number of SHS households to be selected from a village, N is the total number of SHS households of all sizes and ages, and N_h is the total number of SHS households belonging to the hth size group. Data on 25 SHS households was collected from each of the treated villages.

Special attention was given to data collected on those control households that matched treated households with respect to certain observable characteristics. For the set of 2,000 control households, lists of cumulative SHS installations in the villages of the respective upazilas were collected from the POs operating in those areas. The level of SHS concentration by village was assessed by collating the lists from these POs. The two villages with least SHS concentrations were selected as control villages for the two treated villages in the same upazila.[11] Data on 31–32 households was collected from each of the two villages. Since the control villages were located in the same upazila but had fewer SHS installations, one might assume a sizeable overlap between them and the treated villages in terms of observable characteristics, making them ideal control villages. Besides controlling for village-level characteristics, treated households were matched with non-treated households based on household-level observable characteristics.

For the set of 400 control households located in the treated villages, special instructions were given to the enumerators. Data on the 6–7 households that had not adopted the SHS technology were collected from each of the 64 treated villages. Particular attention was given to choice of these households so that they would match the 25 SHS households that had already been interviewed, based on observed socioeconomic characteristics.

Weighting Procedure

For each household in a sample region or population, weight was defined as the number of total similar households (SHS users or non-users), expressed as follows:

(a) weight for SHS users = P_SHS/S_SHS and
(b) weight for SHS non-users = P_NSHS/S_NSHS,

where P_SHS = number of SHS users in a region or population, S_SHS = number of SHS users in the sample for the same region, P_NSHS = number of non-SHS users in a region or population, and S_NSHS = number of non-SHS users in the sample for the same region. These four numbers needed to calculate weight were derived as follows:

P_SHS: Provided by IDCOL at about the same time the survey was conducted.
S_SHS: Generated from the survey data; aggregated from the sample information.
P_NSHS: Derived by subtracting P_SHS for each region by the total number of rural households.
S_NSHS: Obtained from the survey.

The weight variable calculated in this way is not 100 percent accurate, as the variable P_NSHS also includes grid households in rural areas; thus, to obtain a more accurate value for P_NSHS, the number of grid households in a region should be subtracted. That said, in the absence of such information, the estimated weight variable is a reasonably good approximation.

Notes

1. An upazila, comprising 20–30 villages, is the second lowest tier of regional administration in Bangladesh.
2. Land is used as proxy for income; as a measure of household wealth, landholding is fairly stable and more accurate than income.
3. Here price is expressed in terms of Tk per watt-peak. In reality, this change translates into a much larger price difference since, at the time of this study, the SHS capacity range was 20–85 Wp; a price increase of Tk 100 per Wp for a 50 Wp system, for example, implies an increase of Tk 5,000 (US$64).
4. Char areas are small riverine islands formed by sedimentation from silt carried by rivers. Even though char areas are highly unstable, are not easily accessible, and have minimal or no infrastructure, millions of rural Bangladeshis live there because of extreme poverty.
5. This finding is somewhat surprising since Grameen Shakti, a sister organization of the Grameen Bank, is by far the largest installer of SHS units in Bangladesh. Given the

overlap of microcredit programs in village coverage, it is possible that Grameen's effect on SHS adoption is captured by BRAC's presence in the village.

6. Because the sample lacked sufficient households that had adopted SHS in 2012, they were dropped from this trend analysis.

7. This reflects the role of income or wealth on the demand for fuelwood and other non-fuelwood biomass. For example, wealthier households are more likely to consume more fuelwood and adopt SHS, compared to relatively poorer households.

8. By late 2012, the savings in kerosene use from SHS adoption amounted to more than 40 million liters. Direct emissions reduction amounted to more than 240,000 metric tons of CO_2; indirectly, the savings were much larger, given the amount of natural gas that would have been consumed if an equivalent amount of energy had been grid-supplied. Thus, SHS adoption significantly reduces fossil-fuel consumption.

9. A union parishad is the lowest administrative unit of local government in Bangladesh.

10. Although Bogra and Hobiganj are not among the top 16 districts, they are among the top 24, with reasonably high concentrations of SHS installations.

11. Two treated villages were selected in each of the selected upazilas.

References

Asaduzzaman, M., Muhammad Yunus, A. K. Enamul Haque, A. K. M. Abdul Malek Azad, Sharmind Neelormi, and Md. Amir Hossain. 2013. *Power from the Sun: An Evaluation of Institutional Effectiveness and Impact of Solar Home Systems in Bangladesh*. Report submitted to the World Bank, Washington, DC.

BIDS/World Bank (Bangladesh Institute of Development Studies and World Bank). 2012. "Household Survey Data on Impact Evaluation of Solar Home Systems in Bangladesh." Bangladesh Institute of Development Studies and World Bank, Dhaka.

Siegel, J. R., and A. Rahman. 2011. *The Diffusion of Off-Grid Solar Photovoltaic Technology in Rural Bangladesh*. Medford, MA: Center for International Environment and Resource Policy, The Fletcher School, Tufts University.

CHAPTER 5

Welfare Impacts of Household Adoption

The benefits of electricity are well recognized by development practitioners and in the literature (Dinkelman 2011; Khandker, Barnes, and Samad 2012, 2013; Khandker et al. 2014; UNEP 2013). At the household level, the benefits start immediately through lighting, which is the primary use of electricity in most households. They can eventually go much beyond that to include extended study hours for school-going children, knowledge and information access through electronic media (TV and radio), extended hours of operation for income-generating activities, and increased productivity from electrically operated tools and machinery.

Access to electricity can also impart significant health benefits. Replacement of kerosene for lighting reduces household air pollution (HAP), which, over time, poses serious health hazards to women and children, who spend much of their time indoors. Kerosene replacement also reduces emission from carbon dioxide (CO_2), a major greenhouse gas (GHG). Television and other electronic media can impart important health knowledge and information. Finally, access to electricity allows household members to organize their time use in a more productive and rewarding way. This chapter discusses and estimates the benefits of solar home system (SHS) adoption by rural households in Bangladesh, using data from the 2012 survey conducted by the Bangladesh Institute of Development Studies (BIDS) and the World Bank (BIDS/World Bank 2012). Given the limitation of SHS, some benefits achieved through grid electrification may not be possible, yet an exercise to estimate the benefits of solar power can be illuminating.

SHS Gains Expressed as Consumer Surplus

The first and foremost benefit that households enjoy from adopting SHS is the higher quality of lighting. Electricity provides light that is hundreds of times brighter than kerosene-based lighting.[1] Moreover, households get their electric light at a much cheaper cost than kerosene-based light, which becomes obvious when the unit price (i.e., price per kilolumen-hour) is considered.[2] Consumer surplus is a generally accepted way of measuring the savings that results when a household switches from kerosene to electric lighting because of the huge price differential. It is defined as the difference between the amount consumers are willing to pay for a certain quantity of a good or service they consume and the amount they actually pay, assuming that willingness to pay (WTP) is higher; that is, it is the virtual savings by consumers from not paying the extra amount they are willing to pay. While households enjoy consumer surplus when they consume either kerosene or electricity, consumer surplus with the latter is much higher because of its lower unit price, implying a gain in consumer surplus whenever they switch from kerosene to electricity.[3]

Figure 5.1 shows the demand curve for household lighting demand (say, in kilolumen-hours), where P_K and Q_K represent the respective price and quantity of kerosene consumed when the household uses kerosene, and P_E and Q_E are the respective price and quantity of electricity consumed after the household switches from kerosene to SHS. The gain in consumer surplus is derived as follows:

Amount household pays for kerosene, AP_K = area $(B + D) = P_K Q_K$
Amount household pays for electricity, AP_E = area $(D + E) = P_E Q_E$

Figure 5.1 Consumer Surplus from Demand Curve

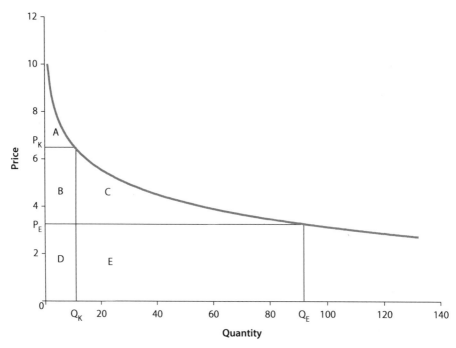

Source: World Bank 2008.
Note: P_E = price of electricity from the grid, P_K = price of kerosene, Q_E = quantity of electricity used from the grid, and Q_K = quantity of kerosene consumed.

Surge in Solar-Powered Homes • http://dx.doi.org/10.1596/978-1-4648-0374-1

Amount household is willing to pay for kerosene, WTP_K = area under demand curve between 0 and Q_K = area (A + B + D)

Amount household is willing to pay for electricity, WTP_E = area under demand curve between 0 and Q_E = area (A + B + D + E + C)

Consumer surplus for kerosene consumption, $CS_K = WTP_K - AP_K$ = area A

Consumer surplus for electricity consumption, $CS_E = WTP_E - AP_E$ = area (A + B + C)

Gain in consumer surplus by switching from kerosene-based lighting to electricity-based lighting = $CS_E - CS_K$ = area (B + C) = $(P_K - P_E)Q_K$ + area C

Using the BIDS/World Bank 2012 survey data, we can determine P_K, Q_K, P_E, and Q_E in a straightforward way; thus, the first term in consumer gain $[(P_K - P_E)Q_K]$ is fairly easy to calculate. However, the shape of the demand curve will determine area C. For example, if the demand curve is a straight line, area C is given by the formula $0.5(P_K - P_E)(Q_E - Q_K)$. However, this formula will overestimate the gain in consumer surplus if the demand curve is convex to the origin, as shown in figure 5.1, and underestimate it if the demand curve is concave. We assume a constant elasticity demand curve (log linear) as suggested in World Bank (2008).[4] A second issue is the assumption that the same demand curve applies to both kerosene users and those who switch to SHS. Ideally, the characteristics that determine demand for lighting energy should be same for these two household groups; but the demand curve may vary between SHS users and non-users (kerosene consumers) for various reasons. For example, because SHS users are likely to be better-off than non-users, their demand curve is more likely shifted upward. Adding to the complexity is that households use kerosene or SHS for additional purposes, making it difficult to determine the lighting-only cost from the total cost of kerosene or electricity. To simplify, we restrict the analysis to households that use kerosene or SHS for lighting only. In addition to aggregate analysis, we calculate consumer surplus by income quintile, assuming households in the same quintile are comparable in terms of energy demand (table 5.1).

For non-SHS households, there is some variation in the price and the quantity of lighting consumed across income quintiles, with the richest households paying the highest price (about Tk 59 per klumen-hr) compared to about Tk 47 per klumen-hr paid by the poorest households. Among SHS households, lighting price does not vary, but the quantity consumed shows a monotonically upward trend from the lowest to the highest quintile, with households in the highest quintile consuming nearly three times as much as those in the fourth quintile. While the consumer-surplus gain is highest for the richest households, it is lowest for this group as a percentage of income. Overall, kerosene lighting is more than 35 times as expensive as SHS-powered electric lighting, and SHS owners consume about 3.5 times as much lighting as non-owners. For lighting use alone, the consumer-surplus gain for SHS adopter households in rural Bangladesh is more than Tk 600 per month.

Consumer surplus is often used to measure project benefits. This can be done by aggregating the gain in consumer surplus for all SHS households.

Table 5.1 Gain in Consumer Surplus from Replacing Kerosene with Electric Lighting

Factor	Income quintile					All households
	1st	2nd	3rd	4th	5th	
Household income, by quintile (Tk/month)	2,670	6,258	9,545	14,046	30,450	11,091
Consumer-surplus component						
Price of kerosene lighting for non-SHS households (Tk/klumen-hr)	47.3	49.2	53.0	51.6	58.7	51.3
Quantity of kerosene lighting consumed by non-SHS households (klumen-hr/month)	7.8	8.7	6.3	7.7	8.0	7.7
Price of electric lighting for SHS households (Tk/klumen-hr)	1.3	1.3	1.4	1.5	1.3	1.4
Quantity of electric lighting consumed by SHS households (klumen-hr/month)	14.9	18.0	21.5	25.0	67.0	26.3
Gain in consumer surplus (Tk/month)	632.1	505.5	458.1	541.3	863.4	607.4
Gain in consumer surplus (% of income)	23.7	8.2	4.8	3.9	2.8	5.5

Source: BIDS/World Bank 2012.
Note: The quantity of lighting is calculated from the lighting hours per day of various appliances (mostly *kupi* and hurricane lamps for non-SHS households and fluorescent tube lights and CFL bulbs for SHS households). The price of lighting is calculated by dividing the monthly energy expenditure by monthly quantity of lighting consumed. Monthly expenditure for SHS lighting is calculated by dividing all costs for SHS adoption (costs of the system and parts replacement over unit lifetime, assumed at 20 years). CFL = compact fluorescent lamp; kL-hr = kilolumen-hours; SHS = solar home system.

More specifically, taking into account the total SHS households in October 2013, at more than 2.4 million (figure 2.1), and applying Tk 607 per month as the gain in consumer surplus for lighting use per household, the aggregate benefit of the SHS project in rural Bangladesh is Tk 1.46 billion per month, equivalent to US$225 million per year. The actual project benefit could be even higher because household use of SHS is not limited to lighting, and each use can have its own gain in consumer surplus. Moreover, SHS installation is happening at a rapid pace, implying that gains in consumer surplus can simply rise at that pace.

Reduction in CO$_2$ Emissions

In developing countries, the populations without electricity use mostly kerosene-based lighting, which poses serious health and safety hazards, along with contributing to global warming. Kerosene lamps are one of the major sources of HAP in developing countries, emitting black carbon, which can cause chronic pulmonary diseases and other respiratory problems. Also, the open flames of kerosene lamps are a fire hazard. In addition, the accumulated effect of kerosene burning on global warming can be enormous. It is estimated that kerosene-based lighting is responsible for emitting 244 million tons of CO$_2$ each year globally. This section uses the BIDS/World Bank 2012 survey data to estimate how much CO$_2$ emissions could be reduced by households switching from kerosene-based lighting to SHS-powered electric lighting.

Household savings in kerosene consumption resulting from SHS adoption can be used to calculate the reduction in CO$_2$ emissions, but it is difficult to calculate because SHS households' past kerosene consumption is not known. Thus, the

Table 5.2 Reduction in CO_2 Emissions from Kerosene Replacement due to SHS Adoption

Factor	CO_2 emission, by income quintile					CO_2 emission for all households
	1st	2nd	3rd	4th	5th	
Non-SHS household	6.2	6.2	6.8	7.5	8.3	6.9
SHS households	1.3	1.3	1.3	1.5	1.6	1.4
Difference in emissions	5.0	4.9	5.5	6.0	6.7	5.5

Source: BIDS/World Bank 2012.
Note: CO_2 = carbon dioxide; SHS = solar home system.

study identified comparable groups of non-SHS users whose kerosene consumption could be used as proxy for SHS users' past kerosene consumption. One way to select comparable SHS and non-SHS households is to select them from the same income group, as was done to estimate consumer surplus.

For both SHS adopter and non-adopter households, there is an upward trend in CO_2 emissions and the differences between the emissions with increases in household income (table 5.2).[5] This implies that the rate of increase in kerosene CO_2 emissions, as household income increases, is higher among non-SHS adopters than adopters. Overall, the reduction in CO_2 emissions from kerosene replacement due to SHS adoption is 5.5 kg a month per household. This translates to nearly 160 million kg in avoided CO_2 emissions each year for all SHS households in Bangladesh—a substantial reduction. Considering that only about 10 percent of people in off-grid areas have adopted SHS, the potential for even greater reductions is quite large.

Econometric Estimation of SHS Benefits

Identifying the broader benefits of SHS adoption is a priority for policy makers; the induced benefits in terms of cost savings from kerosene replacement may not be enough to promote such incentives as subsidy. However, if it is shown that SHS adoption helps to induce other measurable benefits (e.g., higher school enrollment, especially among girls, or improved health as a result of obtaining information from television), then such a program would be worth supporting. How then does one measure the benefits induced from SHS adoption?

To assess how SHS adoption affects household welfare, the study conducted an impact evaluation. However, a possible simultaneity arises as causation may run in the reverse direction (e.g., from household outcome, such as income, to SHS adoption). Indeed, a household's decision to adopt a SHS may depend on income and an array of additional determinants. Also, some determinant variables may be unobservable. For example, households more willing to take risk might be more motivated to purchase a SHS. Such unobservable characteristics may affect the outcome variable. Not controlling for the unobservable characteristics will bias the differences in outcome attributed to SHS adoption. In the econometrics literature, this is commonly referred to as omitted variable or selection bias or endogeneity bias more generally. Endogeneity may also be an issue in the selection of villages where the partner organizations (POs) sell SHS units. The choice of villages for SHS promotion is endogenous in that these villages must be in off-grid areas in order for the POs to sell SHS and receive refinancing and subsidy through Infrastructure Development Company Limited (IDCOL). However, once a PO chooses to operate in a particular village, all households are eligible to purchase a SHS if they so wish.

With cross-sectional data like that used for this study, the difference-in-difference technique or fixed-effects (FE) method, commonly used for impact evaluation, cannot be applied since it cannot control for unobserved determinants that vary over time. The few contending non-experimental methods available for ex-post impact evaluation with cross-sectional data are regression discontinuity design (RDD), instrumental variables (IV), and propensity-score matching (PSM) (Khandker, Koolwal, and Samad 2010). The RDD method is applicable in cases with explicitly specified, exogenous rules for interventions (e.g., microcredit programs in Bangladesh where targeting is based mostly on land ownership); however, it is ruled out in this case because there is no definitive criterion for targeting households for the purpose of SHS adoption. The IV method involves finding suitable instruments that affect SHS adoption but not the outcomes of interest; however, finding such instruments for SHS adoption is difficult.

Due to a lack of suitable instruments, we use PSM to assess the impacts of SHS adoption. The idea is to match program participants with non-participants using observable household and community characteristics that affect SHS adoption, despite the shortcoming of unobservable variable bias. Based on such characteristics, each SHS adopter household is paired with non-adopters in the comparison group with similar probability of SHS adoption. This probability or propensity score is estimated as a function of individual characteristics, typically using a logit or probit model. The program impact is estimated by the difference between the observed mean outcomes of the matched SHS households and non-SHS households (annex 5A).

The welfare effects of SHS adoption may also vary by how long households have used solar panels, the capacity of the unit, or the amount of electricity consumed, as impacts on some outcomes may take time to materialize. Since simple PSM technique, which uses a yes/no or binary intervention variable, cannot be used with continuous variables, such as duration or electricity consumption, we also use a variant of the PSM technique, called *p*-score weighted regression, to estimate the impacts of SHS adoption, duration, and electricity consumption. The *p*-score weighted technique is sometimes preferred as it does not involve sample attrition that may occur with simple PSM because it uses only matched households for impact assessment.[6] Also, alternate estimation methods help us to compare the impacts.

The array of welfare outcomes considered included consumption of kerosene, hours spent collecting fuel, hours spent studying in the evenings, incidence of morbidity among household members, women's decision-making power, and household per capita expenditure. The results of the PSM estimation are discussed below.

Energy Consumption from Alternate Sources and Children's Education

One of the primary uses of SHS is household lighting, which extends the waking and working hours of family members. As previously mentioned, solar lights usually replace kerosene lamps or lanterns of various types, which reduce HAP and permit children to study and ready in the evenings.[7] Extended evening study, in turn, is likely to have positive effects on children's schooling outcomes, such as higher enrollment or years completed. Table 5.3 provides the descriptive statistics of kerosene use and educational outcomes by SHS adoption.

While decreased kerosene consumption due to SHS adoption was discussed in chapter 4, it is reiterated here for convenience. Children's study time and completed schooling years are better in SHS households than in their counterpart non-SHS households, while the difference is not statistically significant in the case of school enrollment. Since descriptive statistics do not imply causality but simply trends, we examine the PSM results. Table 5.4 reports the impacts of SHS adoption on these outcomes from the PSM estimation.

Findings from the PSM estimation are consistent with the descriptive statistics in most cases. Reduction in kerosene consumption due to SHS adoption is 2.4 liters per month, based on simple PSM estimation and 2.3 liters per month, based on p-score weighted regression. Also, kerosene consumption drops by 0.71 liter per month for an additional year of SHS use, and by 0.43 liter per month for an additional kilowatt-hour consumption of electricity from SHS. Children's evening study time increases because of SHS adoption, and it appears to increase more for boys than for girls (by more than 14 minutes versus 7.7–12.4 minutes, respectively).

Based on simple PSM estimation, both children's school enrollment and completed years of schooling increase because of SHS adoption, but not so when based on p-score weighted regression. Among the schooling outcomes, boys' completed schooling years is positively affected by the duration of SHS use, and girls' schooling years by electricity consumption. Overall, it can be said that SHS adoption has positive impacts on children's education and schooling outcomes.

Table 5.3 Household Kerosene Consumption and Children's Educational Outcomes, by SHS Adoption Status

N = 4,000

Outcome variable	SHS households	Non-SHS households	t-statistics of the difference
Kerosene use (liters/month)	0.92	2.84	−23.62**
Evening study time (minutes/day), boys (ages 5–18)	131.3	120.0	2.68**
Evening study time (minutes/day), girls (ages 5–18)	127.3	115.0	2.99**
School enrollment rate, boys (ages 5–18)	0.775	0.732	1.30
School enrollment rate, girls (ages 5–18)	0.820	0.770	1.53
Completed schooling years, boys (ages 5–18)	3.64	3.24	1.70*
Completed schooling years, girls (ages 5–18)	3.81	3.26	2.35**

Source: BIDS/World Bank 2012.

Note: SHS = solar home system.

Significance level: * = 10 percent, ** = 5 percent.

Table 5.4 PSM Estimates of SHS Adoption Impacts on Kerosene Consumption and Children's Educational Outcomes

N = 4,000

Outcome variable	PSM estimates	p-score weighted regression estimates		
	SHS adoption	SHS adoption	Duration of SHS use (years)	Electricity consumption from SHS (kWh/month)
Kerosene use (liters/month)	−2.390** (−40.70)	−2.310** (−36.51)	−0.711** (−12.76)	−0.428** (−11.28)
Evening study time (minutes/day), boys (ages 5–18)	14.7** (8.74)	14.1** (5.36)	4.2** (4.81)	2.7** (5.56)
Evening study time (minutes/day), girls (ages 5–18)	12.4** (7.31)	7.7** (2.77)	2.8** (2.85)	2.6** (4.52)
School enrollment rate, boys (ages 5–18)	0.049** (3.51)	0.030* (1.69)	0.005 (0.81)	0.002 (0.75)
School enrollment rate, girls (ages 5–18)	0.045** (3.37)	0.012 (0.65)	0.004 (0.61)	0.004 (0.90)
Completed schooling years, boys (ages 5–18)	0.493** (4.62)	0.215 (1.40)	0.085* (1.63)	0.042 (1.26)
Completed schooling years, girls (ages 5–18)	0.518** (4.77)	−0.005 (−0.03)	0.084 (1.32)	0.067** (2.03)

Source: BIDS/World Bank 2012.

Note: For PSM estimation, nearest neighbor (NN) matching technique is used in this and subsequent tables. The regressions reported in this and subsequent tables control for all exogenous variables reported in the demand equation, as shown in table 5.3. kWh = kilowatt hour; PSM = propensity-score matching; SHS = solar home system.

Significance level: * = 10 percent, ** = 5 percent.

Income, Expenditure, and Assets

Numerous studies have demonstrated the positive impacts of grid electrification on income and expenditure (Khandker, Barnes, and Samad 2012, 2013; Khandker et al. 2014). These studies show electrification impacts of up to 39 percent on income and 23 percent on expenditure. Because of the limited capacity of SHS, its adoption by households is unlikely to have impacts of such scale. With grid electrification, an important channel of income growth is the use of electric motive power in various tools and machinery. With SHS, such an option is virtually non-existent, at least for the foreseeable future in Bangladesh. Still, SHS adoption can boost income in at least two ways: (a) allowing home-based, income-generating activities to remain open for longer hours in the evenings and (b) giving household members access to news and information through electronic media, such as TV and radio.

Also, SHS makes it possible to charge mobile phones—now ubiquitous in rural villages—at home, which becomes an avenue for generating income. Prior to electrification, mobile phone owners had to commute to designated charger locations. Today, adopter households can earn extra money by charging non-adopters' mobile phones, which saves those residents the expense of frequent commutes. With more income, household expenditure can grow, and in the process, households can acquire assets, particularly durable home goods. From the descriptive statistics for these outcomes, it is obvious that SHS households are better off than their counterpart non-SHS households (table 5.5). Compared to non-adopters, SHS adopters have 38 percent higher income, 33 percent higher expenditure, and twice as much in assets. But how much of these differences can be attributed to SHS adoption? To answer this question, we estimate the impacts of SHS adoption on these outcomes.

SHS adoption indeed has positive impacts on household income and expenditure (table 5.6), although the scale is smaller than that of grid electrification impacts. The impacts are higher on income than expenditure, and are higher based on simple PSM than *p*-score weighted regression. The impacts on expenditure and income are about 4–5 percent and up to 12 percent, respectively. An additional year of SHS adoption increases household per capita income by 2.5 percent and per capita expenditure by about 1.6 percent. Household ownership of domestic goods can increase by 23–27 percent because of SHS adoption. For an additional kilowatt-hour consumption of electricity from SHS, durable home goods can increase by 6 percent. Although the increase in domestic goods does not translate to statistically significant growth in total assets, SHS adopters have twice as much in assets as non-adopters. At the time data were collected for this study, households in rural Bangladesh had been adopting SHS under IDCOL's program for nearly a decade; it appears that is a long enough time for income, expenditure, and assets to be affected by SHS adoption.

Table 5.5 Household Income, Expenditure, and Assets, by SHS Adoption Status
N = 4,000

Outcome variable	SHS households	Non-SHS households	t-statistics of the difference
Per capita income (Tk/month)	3,187.9	2,303.3	5.36**
Per capita expenditure (Tk/month)	2,843.3	2,133.2	8.06**
Domestic durable goods (Tk)	34,071.7	19,745.8	12.86**
Total assets (Tk)	5,798,225.0	2,586,596.0	7.27**

Source: BIDS/World Bank 2012.
Note: SHS = solar home system.
Significance level: * = 10 percent, ** = 5 percent.

Table 5.6 PSM Estimates of Impacts of SHS Adoption on Income, Expenditure, and Assets
N = 4,000

	PSM estimates	p-score weighted regression estimates		
Outcome variable	SHS adoption	SHS adoption	Duration of SHS use (years)	Electricity consumption from SHS (kWh/month)
Log per capita income (Tk/month)	0.123** (2.39)	0.086** (2.25)	0.025* (1.91)	0.030** (3.14)
Log per capita expenditure (Tk/month)	0.051** (2.01)	0.042 (1.49)	0.016* (1.84)	0.017** (2.60)
Log domestic durable goods (Tk)	0.266** (6.80)	0.225** (6.78)	0.077** (6.95)	0.062** (7.87)
Log total asset (Tk)	0.067 (1.00)	0.007 (0.28)	0.006 (0.81)	−0.001 (−0.25)

Source: BIDS/World Bank 2012.
Note: kWh = kilowatt hour; PSM = propensity-score matching; SHS = solar home system.
Significance level: * = 10 percent, ** = 5 percent.

Health and Women's Fertility Behavior

Adoption of SHS or any electrification intervention is expected to lower incidence of illness, particularly that of respiratory diseases or related illnesses and gastrointestinal diseases. These health benefits can come through two channels. First, reduced kerosene use lowers HAP due to kerosene smoke. Second, health-related news and information acquired through electronic media, such as TV, increase hygienic practices on the part of household members. Households that acquire a typical SHS unit often purchase a black-and-white television set. In fact, the presence of electricity may encourage households to buy a TV or radio, which they would not have otherwise done. As a source of education and entertainment, TV offers useful information that can enhance knowledge and awareness of events and activities that are economically, socially, or health-wise beneficial. Furthermore, women gain awareness of reproductive health issues, which can motivate them to change their reproductive behavior on contraception and recent fertility and empower them in household decision-making.[8] It is argued that women and young children benefit the most health-wise from access to electricity. Table 5.7 shows descriptive statistics on the incidence of respiratory and gastrointestinal diseases among women and children, as well as women's fertility outcomes (contraceptive use and recent fertility), by SHS adoption.

Table 5.7 shows no consistent pattern in the incidence of diseases among women and children by SHS adoption, and differences between adopter and non-adopter households are not statistically significant. As for reproductive behavior, contraceptive prevalence is higher among married women of non-adopter households than among those of adopter households, and their difference is statistically significant, which is counter-intuitive. While recent fertility in SHS households is higher than in non-SHS households, the difference between the two groups is not statistically significant, much like the descriptive statistics.

Table 5.7 Health and Women's Fertility Behavior, by SHS Adoption Status
N = 4,000

Outcome variable	SHS households	Non-SHS households	t-statistics of the difference
Incidence of disease in last 12 months			
Respiratory, women (ages 15 and over)	0.086	0.092	−0.58
Gastrointestinal, women (ages 15 and over)	0.090	0.084	0.53
Respiratory, boys (under age 15)	0.132	0.126	0.32
Gastrointestinal, boys (under age 15)	0.092	0.080	0.79
Respiratory, girls (under age 15)	0.118	0.128	−0.54
Gastrointestinal, girls (under age 15)	0.093	0.091	0.18
Contraceptive use, married women (ages 15–49)	0.681	0.788	−4.96**
Recent fertility, married women (ages 15–49)	0.394	0.370	0.82

Source: BIDS/World Bank 2012.
Note: SHS = solar home system.
Significance level: * = 10 percent, ** = 5 percent.

Table 5.8 Health and Women's Fertility Behavior, by TV Ownership in SHS Households

N = 1,600

Outcome variable	Households with TV	Households without TV	t-statistics of the difference
Incidence of disease in last 12 months			
Respiratory, women (ages 15 and over)	0.074	0.095	−2.07**
Gastrointestinal, women (ages 15 and over)	0.068	0.109	−3.90**
Respiratory, boys (under age 15)	0.123	0.137	−0.79
Gastrointestinal, boys (under age 15)	0.071	0.105	−2.28**
Respiratory, girls (under age 15)	0.095	0.132	−2.32**
Gastrointestinal, girls (under age 15)	0.074	0.105	−2.12**
Contraceptive use, married women (ages 15–49)	0.695	0.670	1.13
Recent fertility, married women (ages 15–49)	0.342	0.434	−3.36**

Source: BIDS/World Bank 2012.
Note: SHS = solar home system.
Significance level: * = 10 percent, ** = 5 percent.

We then look at the PSM estimates of the impact of SHS on those outcomes. It appears that SHS adoption alone has yet to make any difference in the health and reproductive outcomes.

To examine whether television makes a difference, we look at the descriptive statistics of the outcomes by TV ownership among SHS households. Results show that, in most cases, health and fertility outcomes vary, and their differences are statistically significant (table 5.8). To investigate whether TV ownership really matters, we look at the PSM estimates of the impacts, this time assuming SHS adoption, along with TV ownership, as the intervention variable.

The results show that SHS adoption alone does not improve health outcomes of women and children and fertility outcomes of women; however, in conjunction with TV ownership, it does matter to those outcomes (table 5.9). TV ownership lowers the incidence of respiratory disease among women by 1.7 percentage points and the incidence of gastrointestinal disease by 3.3 percentage points. It also lowers incidence of both diseases among girls under 15, while, in the case of boys, only incidence of gastrointestinal disease decreases as a result of TV ownership. Among the fertility outcomes, women's recent fertility decreases by 6.3 percentage points, while contraceptive prevalence is unaffected. Unlike the findings of simple PSM estimation, that of *p*-score weighted regression show fewer outcomes being affected by TV ownership. Overall, the finding is interesting as it validates a hypothesis that knowledge and information gathered from television backed by electricity can indeed contribute to welfare enhancement.

Change in Time-Use Patterns

With access to SHS lighting and television, family members' time-use patterns change in ways that lead to longer-term socioeconomic benefits. As previously

Table 5.9 PSM Estimates of Impacts of TV Ownership on Health and Women's Fertility

N = 4,000

Outcome variable	PSM estimates	p-score weighted regression estimates
Incidence of disease in last 12 months		
Respiratory, women (ages 15 and over)	−0.017** (−2.220)	−0.0002 (−0.01)
Gastrointestinal, women (ages 15 and over)	−0.033** (−4.334)	−0.022* (−1.64)
Respiratory, boys (under age 15)	0.006 (0.418)	0.012 (0.67)
Gastrointestinal, boys (under age 15)	−0.023** (−2.104)	−0.001 (−0.07)
Respiratory, girls (under age15)	−0.029** (−2.291)	0.008 (0.37)
Gastrointestinal, girls (under age 15)	−0.026** (−2.339)	−0.036** (−2.78)
Contraceptive use, married women (ages 15–49)	−0.009 (−0.526)	−0.021 (−1.13)
Recent fertility, women (ages 15–49)	−0.063** (−2.921)	−0.015* (−1.90)

Source: BIDS/World Bank 2012.
Note: PSM = propensity-score matching.
Significance level: * = 10 percent, ** = 5 percent.

Table 5.10 Women's Time Use for Selected Activities, by SHS Adoption Status

N = 4,000

Women's time use (hours/day)	SHS households	Non-SHS households	t-statistics of the difference
Fuel collection	0.262	0.413	−4.33*
Study/reading	0.714	0.667	0.34
Tutoring children	0.169	0.097	3.14**
Leisure or rest	2.912	2.767	1.41

Source: BIDS/World Bank 2012.
Note: SHS = solar home system.
Significance level: * = 10 percent, ** = 5 percent.

discussed, better-quality lighting results in school-going children—both girls and boys—studying longer in the evening, which positively impacts educational outcomes. Women with access to higher-quality lighting can manage their household chores at a less hurried pace throughout the day and re-allocate freed-up time to income-generating, educational, and leisure activities. A sense of security from lighting may lead to increased social interactions. Longer waking and working hours may alter time-use patterns of both men and women, but possibly more for women.[9] This subsection examines how SHS adoption can affect women's allocation of time to selected activities.

Women in SHS households spend less time collecting fuel and more time tutoring children than their counterparts in non-SHS households (table 5.10). Women in adopter households also have more time for study and leisure, compared to those in non-adopter households, but the differences are not statistically significant. To discover how much of the differences are due to SHS adoption, we examine the PSM estimates (table 5.11).

Table 5.11 PSM Estimates of Impacts of SHS Adoption on Women's Time Use for Selected Activities

N = 4,000

Women's time use (hours/day)	PSM estimates	p-score weighted regression estimates
Fuel collection	−0.109** (−6.371)	−0.090** (−3.10)
Study/reading	0.155* (1.98)	−0.234 (−1.18)
Tutoring children	0.054** (3.41)	0.008 (0.30)
Leisure or rest	−0.107* (−1.94)	−0.098 (−0.97)

Source: BIDS/World Bank 2012.
Note: PSM = propensity-score matching; SHS = solar home system.
Significance level: * = 10 percent, ** = 5 percent.

The results of both the PSM and *p*-score weighted regression estimates show that SHS adoption results in women spending less time collecting fuel (table 5.11). Specifically, women spend about 0.11 hour less per day on fuel collection, which translates to a weekly time savings of about 46 minutes. Reduced fuel-collection time does not necessarily imply that SHS households consume less fuelwood than their counterpart non-SHS households. As shown in chapter 4, the opposite is the case (table 4.5). A possible explanation for the reduced fuel-collection time is that SHS households buy more fuel and collect less. Indeed, the data show that SHS households purchase nearly 60 percent more fuelwood than non-SHS households. The reduction in fuel-collection time could imply a shift in preference, whereby women value an alternate activity more than fuel collection; for example, according to simple PSM, women spend more time in study/reading and tutoring children as a result of SHS adoption (table 5.11). SHS adoption increases women's study/reading time by 0.36 hour per day, which is equivalent to 65 minutes of reading per week. Somewhat surprisingly, SHS adoption decreases women's leisure time. One possible explanation is that women in SHS households are engaged too much in other activities to have leisure.

Women's Empowerment

SHS adoption can boost women's empowerment by allowing them to acquire knowledge and information through TV and other electronic media. Moreover, women may feel empowered as a result of economic improvement through small income-generating activities that may result from SHS adoption (table 5.12).

In most cases, the empowerment indicators are better for SHS households than non-SHS households; however, the differences in outcomes between the two groups are statistically significant in only four out of seven outcomes. To determine whether such differences can be attributed to SHS adoption, we examine the PSM estimates.

The results show that SHS adoption enhances women's empowerment in numerous ways, including mobility and a wide range of decision-making abilities, which women can do independently (table 5.13). For example, according to simple PSM estimation, women's ability to visit parents' house increases by

Table 5.12 Measures of Women's Empowerment, by SHS Adoption Status
N = 4,000

Women's independent decision-making ability	SHS households	Non-SHS households	t-statistics of the difference
Visiting parents by herself	0.207	0.150	2.58*
Children's issues	0.028	0.017	1.34
Own health issues	0.393	0.319	2.59**
Family issues	0.051	0.030	1.94*
Purchase of own goods	0.174	0.176	−0.10
Purchase of household goods	0.031	0.021	1.13
Family planning	0.181	0.121	2.93**

Source: BIDS/World Bank 2012.
Significance level: * = 10 percent, ** = 5 percent.

Table 5.13 PSM Estimates of Impacts of SHS Adoption on Women's Empowerment
N = 4,000

Women's independent decision-making ability	PSM estimates	p-score weighted regression estimates
Visiting parents by herself	0.032** (2.93)	−0.004 (−0.19)
Children's issues	0.008* (1.97)	0.003* (1.67)
Own health issues	0.039** (2.89)	−0.026 (−1.14)
Family issues	0.015** (2.94)	0.008** (2.02)
Purchase of own goods	0.035** (3.62)	0.012* (1.66)
Purchase of household goods	0.007* (1.66)	0.002 (0.46)
Family planning	0.028** (2.81)	−0.0004 (−0.03)

Source: BIDS/World Bank 2012.
Note: SHS = solar home system.
Significance level: * = 10 percent, ** = 5 percent.

3.2 percentage points because of SHS adoption. According to both estimations, SHS adoption improves women's decision-making on children's and other family issues. Women's decision-making on purchases of own and household goods also improves as a result of SHS adoption. Finally, women's decision-making ability on family-planning issues increases by 2.8 percentage points because of SHS adoption.

Are SHS Households Better Off?

This section attempts to answer the key research question: has adopting solar power caused changes in the behavioral responses of households; that is, are they better off as a result of having adopted SHS? The assessment has revealed several short- and medium-term benefits of SHS adoption. First, there is a significant project benefit in terms of gains in consumer surplus by switching from kerosene-based to SHS-powered electric lighting. Also, kerosene replacement contributes directly to lowering HAP and CO_2 and other GHG emissions. Second, the econometric analysis shows welfare gains due to SHS adoption for a variety of outcome indicators. SHS adoption increases the evening study time of school-going boys and girls by up to 15 minutes and 12 minutes, respectively. Household income, expenditure, and assets appear to grow as a result of SHS adoption. Solar power increases per capita income by 9 to 12 percent, per capita expenditure by 4 to 5 percent, and durable home goods by 23 to 27 percent. Women and children in households with TV among the SHS adopters experience less incidence of gastrointestinal and respiratory diseases, and women's recent fertility also improves in those households. Women's time in SHS households is spent on more fruitful pursuits; less time is spent on fuelwood collection, while more is used for own study or reading and tutoring children. Finally, in SHS adopter households, women's mobility and intra-household decision-making ability seems to improve. However, we do not observe any significant difference in women's income-generating activities as a result of SHS adoption, which is somewhat surprising.

Interestingly, the welfare effects of SHS adoption compare favorably with those of grid electrification. A recent study using 2005 household survey data from rural Bangladesh finds that grid connectivity improves household per capita income by 21 percent and per capita expenditure by 11 percent and the evening study time of boys and girls by about 22 minutes and 12 minutes, respectively (Khandker, Barnes, and Samad 2012).[10] Because of its more limited capacity, utility, and services, the benefits of SHS adoption are not expected to match those of grid connectivity. But such a comparison shows that the benefits achieved tend toward the right direction.

Do the Benefits Outweigh the Costs?

The study calculated the cost-effectiveness of SHS for households using a simple three-step process. First, the extent of benefits generated from a SHS for an average household were quantified. As previously noted, substituting kerosene with solar-powered lighting reduces the average SHS adopter household's kerosene consumption by 2 liters per month; this consumption savings translates to a cost savings for lighting of Tk 160 per month.[11] Second, the income gains induced by SHS adoption are considered. Adopter households accrue an 8.6 percent increase in total income (from table 5.6, using lower of the two impacts), equivalent to a monthly gain of more than Tk 1,500. The cost-savings benefits for lighting (Tk 160) and total expenditure (Tk 1,500) are taken together as a measure of accrued household benefits from SHS adoption, equal to Tk 1,600 per month.

Finally, this benefit is divided by the average monthly household cost on a SHS unit to obtain the benefit-cost ratio. In 2012, the average unit price, including the subsidy and interest, was Tk 26,237. Since a solar panel lasts about 20 years, we add to it the costs of battery (about 30 percent of the system cost lasting about five years), controller (about 3 percent of the system cost lasting about three years), and maintenance and repair (Tk 500 per year after the initial three-year warranty period), and 10 percent additional system costs (covering other spare parts and light bulbs), to obtain the aggregate system cost for 20 years, which totals about Tk 60,000.[12] From this cost, we obtain a monthly cost of Tk 250, resulting in a benefit-cost ratio of 6; that is, the accrued benefit of a SHS unit exceeds its cost by 500 percent.[13] Given the social cost of marketing SHS in terms of grant and subsidy support, the next question is whether the social benefits generated through higher access to electricity via solar power exceed the social cost. Exploring this question is the subject of the next chapter.

Annex 5A: Note on Propensity-Score Matching Technique

PSM requires that participants in any intervention (SHS households in this study) match with "otherwise identical" non-participants (non-SHS households), and quantifies any difference in outcome variables between these two groups. That is, this identification strategy relies on the observable characteristics at village and household levels to match program participants with non-participants. This reduces potential bias due to the non-randomness of program placement (village selection for SHS dissemination) and program participation (SHS adoption).

Using the PSM technique, we essentially construct a statistical comparison group based on a model of the probability of SHS adoption. This helps to estimate the probability of participation, also called propensity score. Given the set of observable covariates, x, potential outcomes are independent of participation so that selection is based solely on observable characteristics and all variables that influence participation and potential outcomes are simultaneously observed (Ravallion 2008). We then find, for each participant, a sample of non-participants that have similar propensity scores. An advantage of the PSM technique over the ordinary least squares (OLS) method is that PSM reduces the number of dimensions on which to match participants and comparison units.

The treatment group and matched comparison group form what is called the *common support*. Any observations from the comparison group that are not within the common support are discarded. If a good number of observations get dropped in a non-random fashion, the estimated impacts may suffer from what is called *sampling bias*. After the propensity score is calculated, different matching criteria can be used to assign participants to non-participants on the basis of the propensity score. This is done by calculating a weight for each matched participant and non-participant set, and the choice of a particular matching technique may affect the estimated impact, average treatment of the treated (ATT), formally expressed as follows:

$$ATT = \sum_{i=1}^{N_T} (y_{ij}^T - \sum_{j=1}^{N_C} W_{ij} y_{ij}^C) / N_T, \tag{5A.1}$$

where y_{ij}^T = outcome for SHS households, y_{ij}^C = outcome for non-SHS households, N_T = number of SHS households, N_c = number of matched non-SHS households, and W_{ij} = associated propensity score–based weight given to non-SHS households in matching with the SHS households.

Below we briefly describe four common matching techniques.

Nearest Neighbor Matching. Nearest neighbor (NN) is one of the most frequently used matching techniques, whereby each treatment unit is matched to the comparison unit with the closest propensity score. One can also choose *n* nearest neighbors and do matching (usually n = 5 is used). Matching can be done with or without replacement. Matching with replacement, for example, means that the same non-participant can be used as a match for different participants.

Caliper/Radius Matching. If the difference in propensity scores for a participant and its closest non-participant neighbor is quite high, NN matching may not provide a good match. This can be avoided by imposing a threshold on the maximum propensity-score distance, called *caliper*. This procedure therefore involves matching with replacement, only among comparison units with propensity scores within a certain range. A higher number of dropped non-participants is likely, however, potentially increasing the chance of sampling bias.

Stratification/Interval Matching. This type of matching first partitions the common support into different strata (or intervals), and then calculates the impact for each interval. Within each interval, the program effect is the mean difference in outcomes between treated and control observations. A weighted average of these interval-impact estimates yields the overall program impact, whereby weight is defined as the share of participants in each interval.

Kernel and Local Linear Matching. With the methods described above there is a possibility that only a small subset of non-participants will ultimately satisfy the criteria to fall within the common support and thus construct the counterfactual outcome. On the other hand, non-parametric matching estimators, such as kernel matching (KM) and local linear matching (LLM), use a weighted average of all non-participants to construct the counterfactual match for each participant.

Besides simple PSM, an alternate implementation of propensity score was tried, which is called *p*-score weighted regression, under the assumption of conditional exogeneity (Hirano, Imbens, and Ridder 2003); *p*-score weighted regression uses OLS to estimate the impacts using a weight variable, where the weight is $1/\hat{P}$ for SHS households and $1/(1-\hat{P})$ for non-SHS households, where *P* is the estimated *p*-score. This specification attempts to control for latent differences across treatment and comparison units that would affect participation in program and resulting outcomes.

Notes

1. Kerosene lighting is provided mainly through wick lamps (locally called *kupi*) or slightly improved hurricane lamps.

2. Lumen is a measure of the brightness of light; simply speaking, it is the amount of light emitted by a source per second. Kilolumen-hour is the amount of light consumed in one hour, measured in thousand lumens.

3. The calculation of consumer surplus demonstrated here is based on the methodology outlined in World Bank (2008).

4. With the assumption of constant elasticity, the area C is given by the formula $C = \frac{K}{\eta+1}(Q_E^{\eta+1} - Q_K^{\eta+1}) - (Q_E - Q_K)P_E$, where η is the elasticity given by the formula $\eta = \frac{\ln(P_K) - \ln(P_K)}{\ln(Q_K) - \ln(Q_K)}$, and K is constant in the demand function, given by $P = KQ^{\eta}$. Details of these derivations can be found in World Bank (2008).

5. It is assumed that 1 liter of kerosene emits 2.5 kg of CO_2.

6. In fact, when many observations are omitted because of the matching requirement, the sample may no longer be representative, and simple propensity-score matching (PSM) is likely to give biased estimates of impacts.

7. While kerosene could be used for cooking, this is a costly alternative to biomass cooking fuel and is seldom used by rural households.

8. Recent fertility is defined as the number of live births during the preceding three years for currently married women ages 15–49.

9. These issues need to be investigated, keeping in mind the seasonal variations in daylight hours.

10. One may compare these returns with the 9–12 percent and 4–5 percent respective increases in income and expenditure because of solar home system (SHS) adoption.

11. This figure is based on a kerosene price of Tk 80 per liter during the data collection for this study.

12. A useful resource to assess the relative costs of various components of SHS, particularly in the context of Bangladesh, is Haque and Das (2013).

13. This is a fairly basic way to estimate costs and benefits as it does not take into account the benefits of carbon dioxide (CO_2) reduction and other non-income benefits or such accounting issues as net present value or internal rate of return; the purpose of this calculation is simply to suggest the potential benefits of SHS adoption compared to its costs.

References

BIDS/World Bank (Bangladesh Institute of Development Studies and World Bank). 2012. "Household Survey Data on Impact Evaluation of Solar Home Systems in Bangladesh." Bangladesh Institute of Development Studies and World Bank, Dhaka.

Dinkelman, Taryn. 2011. "The Effects of Rural Electrification on Employment: New Evidence from South Africa." *American Economic Review* 101 (7): 3078–108.

Haque, S. M. Najmul, and Barun Kumar Das. 2013. "Analysis of Cost, Energy and CO_2 Emission of Solar Home Systems in Bangladesh." *International Journal of Renewable Energy Research* 3 (2): 347–52.

Hirano, Keisuke, Guido Imbens, and Geert Ridder. 2003. "Efficient Estimation of Average Treatment Effects Using the Estimated Propensity Score." *Econometrica* 71 (4): 1161–89.

Khandker, Shahidur R., Gayatri B. Koolwal, and Hussain A. Samad. 2010. *Handbook on Impact Evaluation: Quantitative Methods and Practices.* Washington, DC: World Bank.

Khandker, Shahidur R., Douglas F. Barnes, and Hussain A. Samad. 2012. "The Welfare Impacts of Rural Electrification in Bangladesh." *Energy Journal* 33 (1): 187–206.

―――. 2013. "Welfare Impacts of Rural Electrification: A Panel Data Analysis from Vietnam." *Economic Development and Cultural Change* 61 (3): 659–92.

Khandker, Shahidur R., Hussain A. Samad, Rubaba Ali, and Douglas F. Barnes. 2014. "Who Benefits Most from Rural Electrification? Evidence in India." *Energy Journal* 35 (2): 75–96.

Ravallion, Martin. 2008. "Evaluating Anti-Poverty Programs." In *Handbook of Development Economics.* Volume 4, edited by Paul Schultz and John Strauss. Amsterdam: North-Holland.

UNEP (United Nations Environment Programme). 2013. *Global Trends in Renewable Energy Investment 2103.* Frankfurt School-UNEP Centre/BNEF. http://www.fs-unep -centre.org.

World Bank. 2008. *The Welfare Impact of Rural Electrification: A Reassessment of the Costs and Benefits.* Independent Evaluation Group (IEG) Impact Evaluation. Washington, DC: World Bank.

CHAPTER 6

Market Analysis and Role of the Subsidy

Given the rapid growth in extending solar home system (SHS) technology to off-grid rural Bangladesh, evidence of the substantial welfare benefits that accrue to rural households that adopt SHS, and the future uncertainty of national grid extension in disadvantaged regions of the country, what is the potential market demand for SHS in off-grid areas? To what extent has the subsidy played a role in achieving the current rate of household adoption? Given the size of the current and potential future SHS market, is the subsidy still needed for further expansion in rural off-grid areas? This chapter attempts to answer these questions to better assess the incentives needed for continued market expansion that ensures reliable service to households and financial viability of the partner organizations (POs).

Estimating Market Size

The SHS units are sold either for cash or on credit under certain conditions between Infrastructure Development Company Limited (IDCOL) and the POs and between the POs and household clients. Ultimately, however, household clients demand the systems. As with other goods and services, this demand is price-sensitive, although the price depends on various elements, including the subsidy level and interest rate on credit. Such sensitivity of demand also implies a maximum potential market size under a given set of conditions. Given that a substantial number of SHS units have already been installed, policy makers would like to know about the current and future expansion of SHS as a means of accessing modern energy, albeit in a limited way, to households in off-grid areas. Many socioeconomic and technological factors can influence market expansion; however, certain trends are major drivers (IFC/World Bank 2010). These are:

- *Grid expansion.* Typically, a country's annual grid-expansion rate must be at least 2–4 percent to keep pace with growing population. Bangladesh is far from achieving that rate because of the continuous energy shortages owing to persistent generation problems. When provision of reliable supply is a problem in grid areas, expansion in off-grid areas is a remote possibility. Given this scenario, the SHS market is expected to grow in the foreseeable future, with potential coverage of all off-grid areas.

- *SHS technology.* The past few years have witnessed tremendous innovation in SHS technology, and the trend is accelerating. Technological improvements have resulted in substantial improvements in the quality of system components (e.g., solar panel, battery, and light bulbs), which have prolonged their life cycle. Technological improvements are also permitting more energy-efficient light-emitting diode lights to run on SHS, which requires lesser capacity panels and batteries, thus lowering the cost of SHS. These positive trends are certain to increase SHS market growth.

- *SHS price.* Because of the recent above-mentioned improvements in SHS technology, consumers have continued to enjoy a significant drop in product price. SHS price reduction will be a major driver of future market expansion.

- *Kerosene price.* For rural households without electricity in Bangladesh, kerosene is by far the major lighting fuel. Period price hikes for kerosene are an increasing trend, which is likely to create added demand for SHS, which has seen a consistent price decline.

All of these trends point to significant SHS market expansion in the foreseeable future. To help policy makers understand the potential size of the SHS market and the conditions under which it operates, we apply the willingness-to-pay (WTP) concept. As discussed in chapter 5, WTP can be used to estimate the gains in consumer surplus when households switch from kerosene-based lighting to electric lighting. The concept can also be used to estimate market demand, as shown in the next subsection.

Willingness to Pay

In addition to calculating project benefit, WTP can help to estimate whether a product or service has additional demand. This study applied the WTP concept, using the contingency valuation (CV) method. CV is particularly useful to estimate market demand when the preference of consumers or potential consumers is not revealed. The underlying concept is that people often have hidden preferences for certain goods and services that can be expressed in terms of monetary value (Hoevenagel 1994). One way to do so is to present respondents information on alternate hypothetical products or services with arbitrary prices and ask, in the form of a yes/no question, whether they would purchase them. Various detailed accounts of CV implementation can be found in a wide body of literature (Cameron and James 1986; Cummings, Brookshire, and Schulze 1986; Mitchel and Carson 1989).

In this study, survey questions were designed to elicit answers from SHS adopter and non-adopter households concerning their willingness to purchase SHS panels. Households that already owned a SHS were asked whether they would purchase an additional unit of a certain capacity size, price, and interest rate from a set of choices over a three-year period. Non-adopter households were asked whether, given the benefits of adopting a SHS (i.e., lighting, running a black-and-white TV, and charging a mobile phone), they would be willing to purchase a unit of a certain capacity, price, and interest rate from a set of choices over a three-year period. The choices offered to the survey respondents were randomly selected and uniformly distributed; that is, each choice had the same number of respondents so that the averages would not be biased. When the respondents were asked the question, the survey context was explained to ensure they understood both the question and context well in order to decide whether they believed the choice was worth accepting. For each system capacity, three choices were developed, with the middle one corresponding roughly to the average existing price of the same capacity and the other two being about 25 percent higher and lower (annex 6A).[1]

Based on the CV method, the purchase behavior of the household can be expressed using an equation similar to (4.1), as follows:

$$P_{ij} = \alpha + \beta X_{ij} + \gamma V_j + \lambda S_{ij} + \chi W_{ij} + \varepsilon_{ij}, \qquad (6.1)$$

where P_{ij} represents the household's purchase decision (i.e., whether it would buy the system corresponding to the choice offered [a yes/no variable with a value of 1 when it would buy the system and 0 otherwise]); S_{ij} is the system price; W_{ij} is a dummy variable for the capacity of the household-owned SHS unit (0 for non-SHS households); X_{ij} and V_{ij} are respective household- and village-level exogenous variables (e.g., household head's age and education, household's land assets, village infrastructure, and prices of alternate energy sources); ε_{ij} equals an unobserved random error; and β, γ, λ, and χ are parameters to be determined.

Based on the purchase-behavior results from the CV questions, a probit model similar to the one reported in table 4.3 was used to estimate the probability of purchase at various prices, with the exception that this model also included dummy variables for the capacity of units owned by SHS adopters. Various models for explaining the probability of SHS purchase were tested before arriving at the final probit model (table 6.1).

The signs for most of the explanatory variables are expected. Offered price has a negative effect on purchase probability, as do the capacity dummies; that is, households that already own a SHS are less likely to purchase an additional unit. And SHS households with higher-capacity units are even less likely to purchase another unit, compared to those with lower-capacity units. The price of kerosene, the alternate fuel for lighting, has a positive effect on the purchase of additional SHS units, implying a substitution effect. To summarize, it appears that the results of the probit regression establish the validity of the adopted CV method, implying that the estimated WTP can be used with reasonable confidence for policy modeling.

Surge in Solar-Powered Homes • http://dx.doi.org/10.1596/978-1-4648-0374-1

Table 6.1 Estimates of SHS Purchase Probability for Contingency Valuation
N = 3,851

Explanatory variable	Probit estimates	Marginal effects
Price of offered SHS unit (thousand Tk)	−0.028** (−8.22)	−0.011** (−8.22)
Household SHS capacity ownership (1 = yes, 0 = no)		
20 Wp	−1.298** (−9.89)	−0.464** (−9.89)
40 Wp	−1.683** (−13.03)	−0.546** (−13.03)
50 Wp	−1.803** (−15.41)	−0.569** (−15.41)
65 Wp	−1.931** (−11.19)	−0.577** (−11.19)
Sex of household head (1 = male, 0 = female)	−0.230* (−1.73)	−0.084* (−1.73)
Age of household head (years)	0.003 (1.19)	0.001 (1.19)
Education of household head (years)	0.014 (1.39)	0.005 (1.39)
Log household land asset (decimals)	0.068** (2.47)	0.026** (2.47)
Log household non-land asset (thousand Tk)	0.052** (2.18)	0.020** (2.18)
Village price of fuelwood (Tk/kg)	−0.006 (−0.18)	−0.002 (−0.18)
Village price of kerosene (Tk/liter)	0.057** (2.35)	0.021** (2.35)
Log price of SHS (hundred Tk/Wp)	−0.761** (−2.69)	−0.288** (−2.69)
Pseudo R²	0.256	0.256
Mean and standard deviation of dependent variable	0.080 (0.272)	

Source: BIDS/World Bank 2012.
Note: Regression includes additional control variables as reported in table 4.3; figures in parentheses are t-statistics, except for the last row where the figure is standard deviation. kg = kilogram; SHS = solar home system; Wp = watt-peak.
Significance level: * = 10 percent, ** = 5 percent.

The underlying theory for estimating WTP is beyond the scope of this discussion; however, a substantial body of literature is available on the subject (Choynowski 2002; Gunatilake et al. 2006). Following Gunatilake et al. (2006), we calculate WTP as follows:

$$WTP = -\left(\alpha + \beta * \overline{X_{ij}} + \gamma * \overline{V_j} + \chi * \overline{W_{ij}}\right)/\lambda, \tag{6.2}$$

where $\overline{X_{ij}}, \overline{V_j},$ and $\overline{W_{ij}}$ are the respective sample means of X_{ij}, V_j, and W_{ij}. By subtracting the actual price paid for the SHS unit from the WTP amount, we obtain the consumer surplus for current SHS owners. Based on the probit regression, the predicted probability of purchasing an additional SHS unit is calculated. The findings show that a household's net benefit is highest when it purchases a 20 Wp unit, with a consumer surplus of about Tk 28,900, and lowest when it purchases a 65 Wp unit, with about Tk 7,825 in consumer surplus (table 6.2).[2]

Of greater interest, however, is the probability of current SHS owners purchasing an additional unit, which is a good indicator of market demand. The results show that the predicted probability of buying an additional unit varies little by the unit capacity owned, and is just 12.5 percent for SHS owners overall. However, for SHS non-users, the probability of purchase is 62 percent, which is quite high. For the overall rural off-grid market, the predicted probability of

Table 6.2 Estimated Consumer Surplus by SHS Capacity and Predicted Probability of Purchase by Users and Non-Users

Household type	Consumer surplus (Tk)	Probability of purchase (%)
SHS users, capacity ownership		
20 Wp	28,896.9	14.0
40 Wp	16,677.3	10.5
50 Wp	9,226.4	12.5
65 Wp	7,825.1	14.7
All SHS users	14,771.0	12.5
All SHS non-users	n.a.	62.0
All households	5,024.6	58.4

Source: BIDS/World Bank 2012.
Note: n.a. = not applicable; SHS = solar home system; Wp = watt-peak.

Table 6.3 Elasticity of SHS Demand at Various Prices

Average SHS unit price (thousand Tk)	Price elasticity at sample mean
22.4 (initial price)	1.06
20.2	0.93
17.9	0.80
15.7	0.68
13.4	0.57
11.2	0.46

Source: BIDS/World Bank 2012.
Note: SHS = solar home system.

purchase is more than 58 percent, which is clearly indicative of an enormous market potential (table 6.2).

Elasticity of Demand

Results of the probit estimates reported in table 6.1 show that household purchase behavior is sensitive to the offered price of the SHS unit. This means that the probit estimate can be used to estimate elasticity of SHS demand. Using equation (6.1), we calculate the price elasticity of demand as follows:

$$PED = \lambda * \bar{S}_{ij}/\hat{P}_{ij}, \tag{6.3}$$

where $\bar{S}_{ij}$ is the average offered price, and $\hat{P}_{ij}$ is the predicted probability of purchase. We use simulation to estimate the sensitivity of price elasticity. Specifically, we lower the price by 10 percentage points at a time and predict price elasticity until reaching a price that is 50 percent lower than the initial price. Since the price of the SHS unit will decline continuously in the future, this exercise gives an idea of how consumers and providers might react to the price decline (table 6.3).

Currently, price elasticity of SHS demand is estimated at slightly more than 1 on average, which means it is elastic (table 6.3). However, future price drop is likely to make demand increasingly inelastic, suggesting that the POs will tend to raise unit prices if there is an opportunity. Under such conditions, IDCOL should ensure market competition and the entrance of new POs into the market each year.

Market Size

From the WTP estimation, we observe that market demand for SHS is quite high, with a 58 percent predicted probability of purchasing a new unit. At the time of the study survey, about 1.3 million of Bangladesh's 15 million off-grid rural households had adopted a SHS, leaving 13.7 million households as potential customers. Applying the 58 percent predicted probability of purchasing a new system to this population yields 7.95, meaning that the market size is more than five times larger than the current customer base. But what happens to the market size when the SHS unit price declines, which will occur in the future. Again, we use simulation to estimate the predicted probability of system purchase at lower prices.

As the SHS price drops by 50 percent (from Tk 22,400 to Tk 11,200), the probability of purchasing a unit rises from 58 percent to 67 percent; as a result, the number of SHS units that would potentially be sold increases from 8 million to 9.3 million (figure 6.1). These findings do not consider that SHS demand is likely to fall if grid electricity coverage increases in currently off-grid areas or alternate sources of renewable-energy technology, such as mini- or micro-grid, enter areas where SHS is marketed. Conversely, SHS might penetrate into grid-based areas as more grid-connected households want to adopt a unit as a backup power supply. Thus, this study's projection may be overestimated or underestimated, depending on what happens in the future. As such, these results should be accepted with some caution.

Figure 6.1 Trends in Probability of Purchase and Potential Sales with Price Drop

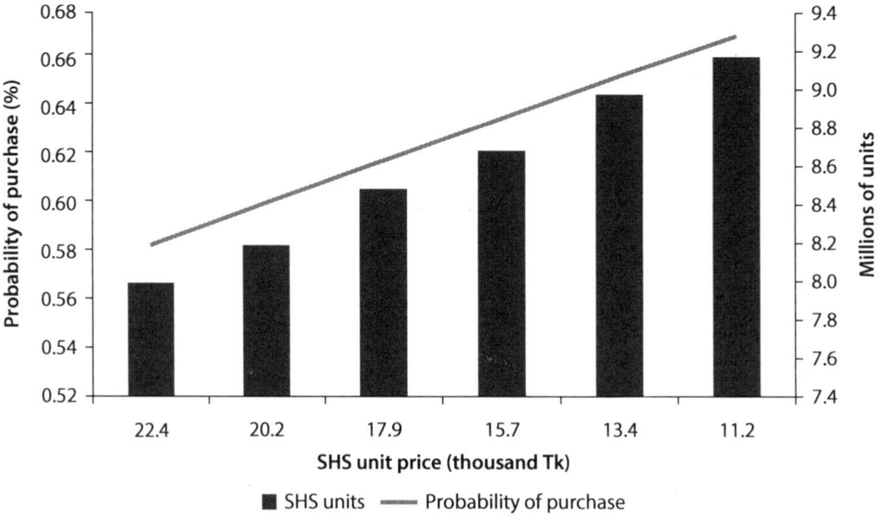

Source: BIDS/World Bank 2012.
Note: SHS = solar home system.

Subsidy Impact on Household Welfare

IDCOL's grant and subsidized loan policies, introduced in 2003, have been instrumental in shifting SHS market demand by allowing the entry of smaller nongovernmental organization POs, even though the extent of the grant per unit has declined over time. With both PO competition and increased market demand, the SHS unit price has fallen, but the subsidy has declined at a faster pace; in 2004, it accounted for about one-quarter of the SHS unit price, compared to less than 10 percent by 2012 (figure 6.2).

The time trend in price and subsidy per watt-peak again shows that the subsidy has declined more than the price (figure 6.3).[3] Importantly, the subsidy-adjusted price also declined thanks to ongoing technological advances in solar panel. The overall implication of these findings is a continued increase in SHS demand.[4]

Financing Mechanism of SHS: A Closer Look

As demonstrated in chapter 2, IDCOL's strategy of incentives for the POs to promote SHS in off-grid rural areas includes (a) a subsidy of US$25 per unit sold, irrespective of system capacity and unit price; (b) 10 percent down payment of the unit price (after deduction of the subsidy) by the household client; (c) refinancing of 80 percent of the amount left after deduction of the subsidy and household client down payment from IDCOL to the PO (at 6 percent interest rate over six years); and (d) loans to household clients for the amount left after deduction of the subsidy and down payment (from PO to household at 12–15 percent interest rate over two-to-three years).[5] Below we consider the implications of changing various incentive structures of the SHS financing mechanism, using Tk 24,585

Figure 6.2 Trend in Subsidy as Percent of SHS Unit Price, 2004–12

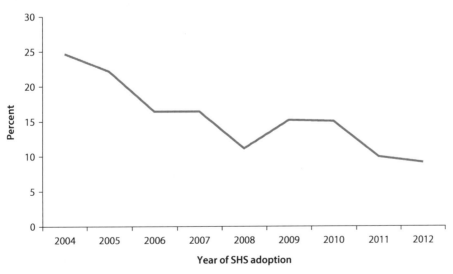

Source: BIDS/World Bank 2012.
Note: SHS = solar home system.

Figure 6.3 Trends in SHS Price and Subsidy, 2004–12

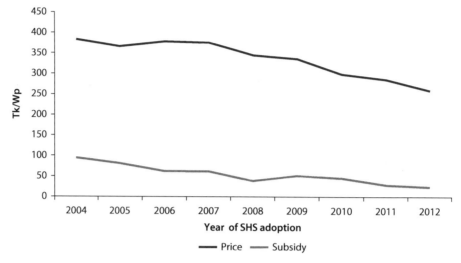

Source: BIDS/World Bank 2012.
Note: SHS = solar home system; Wp = watt-peak.

as the unit price (the sample average before credit interest) and US$25 as the current subsidy amount for a system. More specifically, a simulation exercise is carried out under various scenarios to observe the impact of incentive changes on the total household payment, PO resources, and SHS unit prices.

Grant Subsidy Withdrawal

Table 6.4 demonstrates that the gradual withdrawal of the grant subsidy from its current level of US$25 results in an increase for every cost element. In a scenario of total subsidy withdrawal (last row, table 6.4), the total household payment increases by Tk 7,966, which is 9 percent more than the current total payment. In this scenario, households must make a larger down payment, meaning that IDCOL must allocate more resources as part of its 80 percent commitment; the POs, in turn, must allocate more resources to finance the credit. Overall, this situation implies a 30 percent increase in payment for every dollar of grants withdrawn (last column, table 6.4).

Reducing IDCOL Loan to POs

Table 6.5 demonstrates that reducing IDCOL's loan commitment would place a heavy financial burden on the POs. While reducing the current 80 percent loan financing would have no impact on a household's total payment, the POs would need to find additional resources to ensure sufficient funds for household loan financing, and their rate of return would be reduced. A complete withdrawal of IDCOL's financing would require the POs to increase their investment fivefold, while the rate of their return would drop to one-third of the current figure.

It can be argued that these two added burdens on PO resources might sway them toward selling SHS units on a cash-on-delivery basis. And perhaps

Table 6.4 Effects of Gradual Subsidy Withdrawal

SHS unit price (Tk)	Subsidy (US$)	Take-home price (Tk)	Down payment (Tk)	Loan amount (Tk)	Interest payment (Tk)	Total payment (Tk)	Subsidy reduction (Tk)	Extra household payment due to subsidy reduction (Tk)	Extra payment/Tk of subsidy withdrawal (Tk)
24,585	25	22,560	2,256	20,304	7,309	29,869	0	0	n.a.
24,585	20	22,965	2,297	20,669	7,441	30,406	405	536	1.3
24,585	10	23,775	2,378	21,398	7,703	31,478	1,215	1,609	1.3
24,585	0	24,585	2,459	22,127	7,966	32,551	2,025	2,681	1.3

Source: BIDS/World Bank 2012.
Note: The conversion rate used is US$1 = Tk 81 (the rate at the time of the study). Elements in the first row illustrate the current scenario, while those in subsequent rows show the impacts of gradual subsidy withdrawal (column 2); n.a. = not applicable.

Table 6.5 Impact of Reducing IDCOL's Share of Financing on POs

Loan by PO to household (Tk)	IDCOL's share of finance (%)	Loan by IDCOL to PO (Tk)	POs' own investment (Tk)	POs' return (Tk/year)	POs' rate of return (%)	Current household interest charged (%)	Adjusted interest charged to recover PO loss (%)
20,304	80	16,243	4,061	1,462	36.0	12	12
20,304	60	12,182	8,122	1,706	21.0	12	18
20,304	40	8,122	12,182	1,949	16.0	12	24
20,304	20	4,061	16,243	2,193	13.5	12	30
20,304	10	2,030	18,274	2,315	12.7	12	33
20,304	0	0	20,304	2,436	12.0	12	36

Source: BIDS/World Bank 2012.
Note: The conversion rate used is US$1 = Tk 81 (the rate at the time of the study); elements in the first row illustrate the current scenario, while those in subsequent rows show the impacts of reducing IDCOL's financing (column 2). IDCOL = Infrastructure Development Company Limited; PO = partner organization.

smaller POs would find it difficult to sell the units because of the large investment requirement. Eventually, the market might be dominated by a few large players. Moreover, as their rate of return would fall (from 36 percent to 12 percent), along with financing withdrawal (from 80 percent to zero), the POs would likely recover this loss by increasing the interest rate charged to household clients (last column, table 6.5). Every 10 percent reduction in IDCOL's share of financing would result in a 3 percent adjusted increase in the interest rate. For the POs to retain their rate of return on resources, the interest rate would reach 36 percent by the time IDCOL financing is completely withdrawn, which is a threefold increase from the current rate.

Extending Duration of Household Loan

Currently, IDCOL provides loans to the POs over a six-year period, while households have a three-year repayment period. This means that, as households gradually repay their loans, the POs likely re-circulate this money as loans to others; for each loan a household is given, the PO uses the fund for two such loans and retains additional interest income.

Results of the study simulations show that extending the duration of household loans could negatively impact both household clients and the POs, ultimately resulting in higher unit prices (table 6.6). Extending the household loan from three to six years while retaining the 12 percent interest rate would increase the total household loan payment from Tk 27,613 to Tk 34,923—more than 26 percent of the original loan payment over three years.

Lowering Household Interest Rate

Simulations show that lowering the interest rate on household loans from 12 percent would negatively affect the POs (table 6.7). At the current 12 percent rate, the POs' average rate of return is 36 percent. If IDCOL were to regulate the household interest rate at 8 percent, the POs' rate of return would fall to 16 percent, which would likely result in the POs demanding a corresponding reduction in IDCOL's interest rate. For every percentage point decline in the household interest rate, the POs would experience a 5 percentage point drop in rate of return; to maintain an equitable rate of return, they would expect IDCOL to lower its interest rate by 1.25 percentage points (last column, table 6.7). If the household interest rate were lowered to 6 percent, the same as the rate charged by IDCOL, then the adjusted PO rate would show an invalid negative value; in this case, the POs' loss in rate of return would be too high to repair, meaning that the overall scheme would no longer be profitable for them.

Table 6.6 Impact of Extending Household Loan Duration

Loan amount (Tk)	Loan repayment period (years)	Total household interest paid (Tk)	Total household payment (Tk)
20,304	3	7,309	27,613
20,304	4	9,746	30,050
20,304	5	12,182	32,486
20,304	6	14,619	34,923

Source: BIDS/World Bank 2012.
Note: The conversion rate used is US$1 = Tk 81 (the rate at the time of the study); elements in the first row illustrate the current scenario, while those in subsequent rows show the impacts of extending the duration of the loan to households (column 2).

Table 6.7 Impact of IDCOL Lowering Interest Rate on Household Loans

Household loan amount (Tk)	Interest rate (%)	Interest paid by household (Tk)	Total cost to household (Tk)	POs' rate of return (%)	Adjusted IDCOL interest rate to recover POs' loss (%)
20,304	12	7,309.44	27,613	36	6.0
20,304	10	6,091.2	26,395	26	3.5
20,304	8	4,872.96	25,177	16	1.0
20,304	6	3,654.72	23,959	6	−1.5

Source: BIDS/World Bank 2012.
Note: The conversion rate used is US$1 = Tk 81 (the rate at the time of the study).

Raising Household Interest Rate

An increase, rather than a decrease, in the interest rate of customer loans may be a more likely scenario since the current 12 percent rate is below the market rate charged by commercial banks. We calculate the cost increase to households resulting from increasing the interest rate to 16 percent (the rate charged by commercial banks), 20 percent (the rate charged by Grameen Bank, Bangladesh's largest microcredit lender), and 27 percent (the maximum rate charged by microcredit lenders in Bangladesh) (table 6.8). The findings show that customers charged the commercial bank rate (16 percent) would pay 8 percent more for the system than what they now pay. Increasing the rate to 20 percent and 27 percent, would increase the system cost by 16 percent and 31 percent, respectively. By charging higher interest rates, the POs would benefit substantially. For example, charging the commercial bank rate of 16 percent would raise their rate of return by 20 percent (from 36 percent to 56 percent). Charging the maximum rate charged by microcredit lenders in Bangladesh would cause their rate of return to soar to 111 percent.

Raising Household Interest Rate and Withdrawing Grant Subsidy

Next we consider the impact of raising the interest rate on loans to household customers in conjunction with grant subsidy withdrawal. This scenario is even more probable than the last since the grant subsidy is being reduced continuously, and indeed is expected to be eliminated completely in the near future. In such a scenario, the unit price paid by households would rise by US$25 (the level of the grant subsidy at the time of the survey) or Tk 26,610 (table 6.9). In this case, the

Table 6.8 Effects of Raising Interest Rates Charged to Customers

Unit price after grant subsidy	Household loan amount (Tk)	Interest rate (%)	Interest paid by household (Tk)	Total cost to household (Tk)	Cost increase due to interest-rate rise (%)	POs' rate of return (%)
24,585	22,127	12	7,966	32,551	0	36
24,585	22,127	16	10,621	35,206	8	56
24,585	22,127	20	13,276	37,861	16	76
24,585	22,127	27	17,922	42,507	31	111

Source: BIDS/World Bank 2012.
Note: The conversion rate used is US$1 = Tk 81 (the rate at the time of the study). PO = partner organization.

Table 6.9 Effects of Raising Household Interest Rates and Withdrawing Subsidy

Unit price (Tk)	Household loan amount (Tk)	Interest rate (%)	Interest paid by household (Tk)	Total cost to household (Tk)	Cost increase due to interest-rate rise (%)	POs' rate of return (%)
26,610	23,949	12	8,622	35,232	8	36
26,610	23,949	16	11,496	38,106	17	56
26,610	23,949	20	14,369	40,979	26	76
26,610	23,949	27	19,399	46,009	41	111

Source: BIDS/World Bank 2012.
Note: The conversion rate used is US$1 = Tk 81 (the rate at the time of the study). PO = partner organization.

household's total cost would increase by 8 percent over the current total payment of Tk 35,232, even if the interest rate were kept unchanged at 12 percent. In the extreme case of increasing the interest rate to 27 percent, the household's total payment would increase by 41 percent, which might not be sustainable.

Comparison of Impacts from Withdrawing Incentives

Figure 6.4 allows one to visually compare the effects of the various scenarios for withdrawing incentives. For example, household loan payments increase in two scenarios: (a) withdrawing the subsidy and (b) increasing the loan term to six years. PO resources are also stretched in two scenarios: (a) withdrawing the subsidy and (b) reducing IDCOL financing. The simulations show that withdrawing all incentives from the SHS product-delivery mechanism in rural Bangladesh would be unsustainable, with the greatest impact suffered by household clients. It can be argued that household interest rates would rise from 12 percent to a minimum of 36 percent in order to main the POs' current rate of return. Without incentives, the total unit price paid by households would rise from the current level of Tk 24,207 to nearly Tk 26,893; the same result would occur with withdrawal of grants. In addition, the POs' rate of return would fall to only 12 percent in three years, which is unsustainable. Alternatively, the household loan duration could be extended to six years, in which case the household unit price after loan repayment would rise to Tk 33,475, without any increase in the rate of return.

Simulation results show that changes in IDCOL's role would impact both the household price paid for the SHS unit and interest rate, but the impacts would differ (figure 6.5). For example, withdrawal of grant elements would result in a 16 percent price rise. In this case, the POs would retain their current rate of return without any expected impact on the household interest rate charged. If IDCOL withdrew the loans it provides the POs, there would be no expected

Figure 6.4 Comparison of Effects from Incentives Withdrawal (Predicted Household Payment)

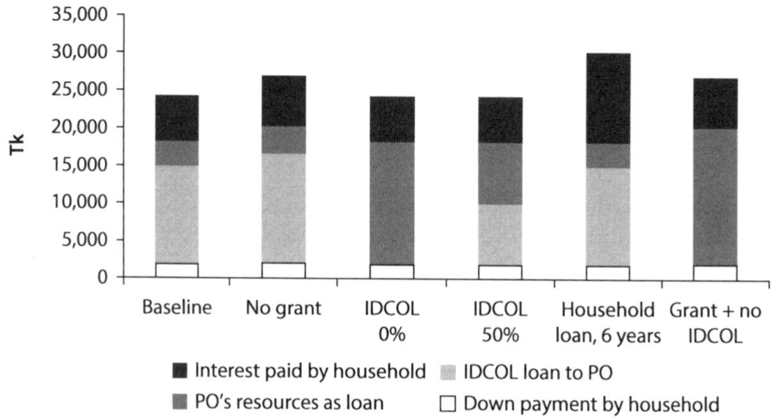

Source: BIDS/World Bank 2012.
Note: IDCOL = Infrastructure Development Company Limited; PO = partner organization.

Figure 6.5 Impact of Changes in IDCOL's Role (Predicted Price of SHS Unit, Interest Rate, and Household Payment)

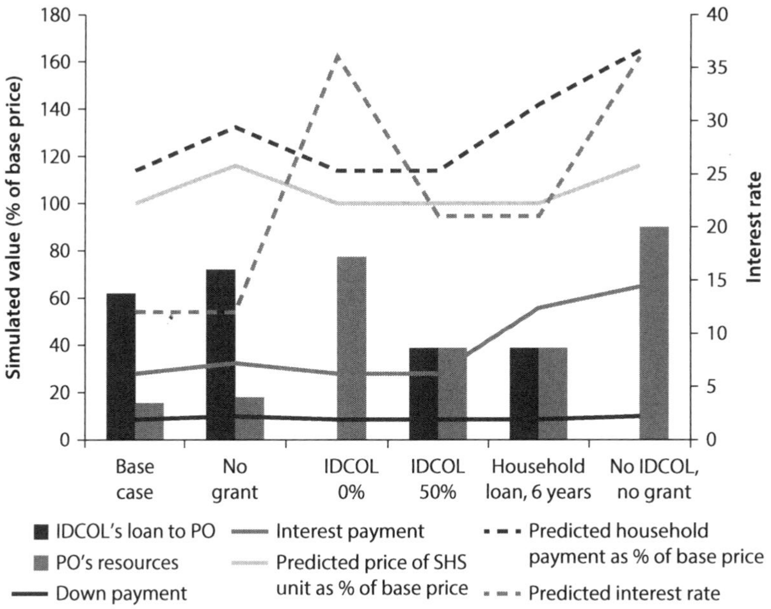

Source: BIDS/World Bank 2012.
Note: The simulation results are for a 20 Wp unit. IDCOL = Infrastructure Development Company Limited; PO = partner organization; SHS = solar home system; Wp = watt-peak.

impact on unit price, but the POs would need greater resources to finance the purchase; at the same rate of interest, the rate of return on investment would fall sharply. To recover this loss, the household interest rate would have to be increased to 36 percent. If IDCOL increased the number of years of the household loan, the impact would be even worse, affecting both the interest rate and final household payment, with no impact on the take-home unit price.

Is the Subsidy Worth Continuing?

As previously discussed, SHS adoption has multiple welfare benefits for households. By providing electricity for lighting, it increases evening study hours for girls and boys. It reduces women's fuel-collection time, and leads to more purchases of television sets; through increased access to media, women's decision-making power is strengthened. Through kerosene replacement, the risk of respiratory disease is reduced. SHS adoption also increases per capita expenditure.

The subsidy that has been used to promote market demand has declined over time. In 2004, it represented 25 percent of the SHS unit price (Tk per watt-peak), but by 2012 accounted for less than 10 percent of the unit price. Over the years, the estimated donor subsidy for SHS expansion has averaged Tk 1,780 per unit of solar panel. Assuming a 10-year life cycle for the SHS unit, the social subsidy to support SHS expansion amounts to only Tk 15 a month per household.

Because kerosene is also subsidized in Bangladesh, one way to estimate the social benefits of solar power may be to compare the savings of the subsidy withdrawn with savings from kerosene replacement. Given that the government provides Tk 12 per liter subsidy to kerosene and that SHS adoption reduces a household's kerosene consumption by 2 liters a month, the benefit to society from not subsidizing kerosene amounts to Tk 24 per month per household. Evaluating the monthly social benefit of Tk 24 against the social cost of Tk 15 means that the social benefit—through kerosene replacement alone—exceeds the social cost by about 60 percent. Other social benefits include cost savings from improved health and better education, as well as the empowerment of women in household decision-making. Thus, continuing the subsidy in SHS program for future expansion may be a good idea. At the same time, it is also true that SHS expansion has been continuously on the rise despite subsidy reduction, implying that the savings in kerosene subsidy would have happened anyway.

Grant subsidy is not the only subsidy in the SHS financing scheme, as the interest rate charged to the customers by the POs (12 percent) is also subsidized because it is below the commercial bank rate (16 percent). Thus, the role of interest subsidy, as well as that of the grant subsidy, must be examined to assess the sustainability of growth in SHS adoption. From the findings shown in tables 6.8 and 6.9, we have already seen that raising the interest rate would cost households more than what they currently pay to own a SHS unit. Raising the interest rate alone would increase the total cost by up to 31 percent; the interest rate hike, combined with complete subsidy withdrawal, would increase the cost by as much as 41 percent.

How likely are current non-users to adopt a SHS unit in such situations? To address this issue, we again use the WTP concept. Applying the CV technique for a sample restricted to SHS non-users, we estimate what non-users would be willing to pay for a new system and compare it with both the current market price of the SHS and the projected prices with the interest rate hike and subsidy withdrawal (tables 6.8 and 6.9).

Even if the grant subsidy were withdrawn completely, non-users would continue to buy SHS units since the willingness-to-buy price (Tk 35,569) is higher than the projected cost after subsidy withdrawal (Tk 35,480) (table 6.10).

Table 6.10 Comparison of Willingness-to-Pay Price and Projected Price with Interest Rate Increase and Subsidy Withdrawal

Interest rate (%)	Amount non-users willing to pay (Tk)	Current system price (Tk)	Projected cost with subsidy withdrawal only (Tk)	Projected cost with interest rate increases only (Tk)	Projected cost with subsidy withdrawal and interest rate increase (Tk)
12[a]	35,569	32,551	35,480	32,551	35,480
16	35,569	32,551	n.a.	35,155	38,084
20	35,569	32,551	n.a.	37,759	41,014
27	35,569	32,551	n.a.	42,641	45,896

Source: BIDS/World Bank 2012.
Note: Shaded cells indicate that projected cost exceeds the amount non-users are willing to pay; n.a. = not applicable.
a. The current interest rate.

This finding is not surprising as we have observed continuous growth in SHS adoption despite the steady decline in the subsidy amount. Also, growth in SHS adoption would not be impeded, even if the POs began to charge interest at the 16 percent commercial bank rate. However, if the POs were to raise the interest rate to 20 percent (that charged by Grameen Bank) or 27 percent (the maximum charged by Bangladesh's microcredit lenders), the projected total system cost would exceed the amount that non-users would be willing to pay. In such a scenario, the SHS adoption rate might not grow as much. The afford-ability of non-user households would be stretched again if the interest rate hike were accompanied by withdrawal of the grant subsidy. Summing up, while the grant subsidy can be eliminated without affecting the growing SHS market demand, raising the interest rate of credit purchase may reduce market demand substantially.

Annex 6A: Contingency Valuation Table Used to Calculate Willingness to Pay

Choice	Watt-peak (a)	Price/ unit (Tk) (b)	Down payment (Tk) (c)	Credit (Tk) (d)	Interest (Tk) (e)	Total price (Tk) (f)	Monthly installment, three-year loan (Tk) (g)	Usage (h)
1	20	9,375	1,406.25	7,968.75	478.13	9,853.13	274	2 lights
2	20	12,500	1,875.00	10,625.00	637.50	13,137.50	365	2 lights
3	20	15,625	2,343.75	13,281.25	796.88	16,421.88	457	2 lights
4	40	16,500	2,475.00	14,025.00	841.50	17,341.50	482	2 lights, TV
5	40	22,000	3,300.00	18,700.00	1,122.00	23,122.00	643	2 lights, TV
6	40	27,500	4,125.00	23,375.00	1,402.50	28,902.50	803	2 lights, TV
7	50	21,375	3,206.25	18,168.75	1,090.13	22,465.13	625	4 lights, TV
8	50	28,500	4,275.00	24,225.00	1,453.50	29,953.50	833	4 lights, TV
9	50	35,625	5,343.75	30,281.25	1,816.88	37,441.88	1,041	4 lights, TV
10	65	26,250	3,937.50	22,312.50	1,338.75	27,588.75	767	5 lights, TV
11	65	35,000	5,250.00	29,750.00	1,785.00	36,785.00	1,022	5 lights, TV
12	65	43,750	6,562.50	37,187.50	2,231.25	45,981.25	1,278	5 lights, TV

Source: BIDS/World Bank 2012.
Note: CV questions are administered in sequence (a–h). The first SHS household is offered choice 1, the second is offered choice 2, and so on, until the thirteenth household is reached, which is again offered choice 1. The same sequence is followed for non-SHS households. CV = contingency valuation; SHS = solar home system.

Notes

1. The capacity choices did not include 75 Wp; thus, households that owned such units were excluded.

2. Currently, off-grid rural households in Bangladesh are increasingly opting for 20 Wp SHS units.

3. The price per watt-peak is calculated by dividing the system price as offered to a rural customer by the capacity of the system adopted. The subsidy, however, has remained the same regardless of system capacity, varying only by year of adoption. Thus, the

subsidy per watt-peak is calculated by dividing the subsidy in the year of adoption by the system capacity adopted by the household customer. Since rural households adopted systems in different years, the price and subsidy have been adjusted by annual consumer price index, with base year 2000, to make the figures compatible across years.

4. In addition to the price decline, rising awareness about the benefits of electricity and SHS ownership have contributed to increasing SHS demand, for which IDCOL and its POs deserve credit.

5. One should note that some details of these arrangements, in place at the time of the study survey, have since changed; for example, the grant subsidy is now US$20 for system capacities below 30 Wp.

References

BIDS/World Bank (Bangladesh Institute of Development Studies and World Bank). 2012. "Household Survey Data on Impact Evaluation of Solar Home Systems in Bangladesh." Bangladesh Institute of Development Studies and World Bank, Dhaka.

Cameron, T. A., and M. D. James. 1986. "Efficient Estimation Methods of 'Closed-Ended' Contingent Valuation Surveys." Working Paper 404, Department of Economics University of California, Los Angeles.

Choynowski, Peter. 2002. *Measuring Willingness to Pay for Electricity*. ERD Technical Note 3. Manila: Asian Development Bank.

Cummings, R. G., S. Brookshire, and W. D. Schulze. 1986. *Valuing Environmental Goods: An Assessment of the Contingent Valuation Method*. Totowa, NJ: Rowman and Allanheld.

Gunatilake, H., J. Yang, S. Pattanayak, and C. Berg. 2006. *Willingness-to-Pay and Design of Water Supply and Sanitation Projects: A Case Study*. ERD Technical Note 19. Manila: Asian Development Bank.

Hoevenagel. R. 1994. "An Assessment of the Contingent Valuation." In *Valuing the Environment: Methodological and Measurement Issues*, edited by E. Pething. Dordrecht, the Netherlands: Kluwer Academic Publishers.

IFC (International Finance Corporation)/World Bank. 2010. *Solar Lighting for the Base of the Pyramid: Overview of an Emerging Market*. Lighting Africa, a Joint Initiative from IFC and the World Bank. http://www.lightingafrica.org.

Mitchel, R. C., and R. T. Carson. 1989. *Using Surveys to Value Public Goods: The Contingent Valuation Method*. New York City: Resources for the Future.

CHAPTER 7

Developing a Sustainable Market

The drivers of consumer demand that influence the development of a robust, regulated solar home system (SHS) market chain in rural Bangladesh work together with key supply-side issues. These include Infrastructure Development Company Limited's (IDCOL's) efforts at transitioning the program toward commercial financing without the need for capital buy-down grants and concessional financing, ensuring reliable after-sales service by the partner organizations (POs), and harnessing possible technological developments in order to cater to consumer needs. This chapter reviews the regulatory functions provided by IDCOL in support of developing a sustainable market for quality SHS in Bangladesh. It also highlights the technical/quality issues that emerged from the 2012 household survey conducted by the Bangladesh Institute of Development Studies (BIDS) and the World Bank and program-level data that would need to be addressed to ensure a sustainable SHS market in the long run (BIDS/World Bank 2012).

Program Incentives and PO Performance

Under the current IDCOL program intervention, the POs are offered the following incentives to ensure that households can afford to buy the SHS units available in the rural market:

- *Buy-down grant.* To help reduce costs to consumers, IDCOL provides a capital buy-down grant (currently US$20 per unit for system capacities below 30 watt-peak [Wp]). Also, smaller POs are given an institutional development grant (currently $3 per system) to help them extend their reach in remote areas. These grants are released after the systems are installed. When the program first started in 2003, the capital buy-down grant was $70 per system, and the PO institutional development grant was $20 per system. As the program expanded, the grant element was reduced, with a goal of eventually withdrawing all grant support (table 7.1).

- *Refinancing of customer credit.* The POs offer households microcredit for purchasing SHS. Households make a down payment of 10–15 percent of the

Table 7.1 Capital Buy-Down Grant (Subsidy) and Institutional Development Grant over Time
US dollars

| | IDCOL program year | | | | | | |
Grant element	*2003*	*2004–05*	*2006–07*	*2008–09*	*2010–11*	*2012*	*2013–14*
Capital buy-down grant	70	55	40	40	25	25	20[a]
Institutional development grant	20	15	10	5	3	0	0

Source: IDCOL (Infrastructure Development Company Limited).
Note: The institutional grant for existing POs was discontinued after 2012. PO = partner organization; SHS = solar home system; Wp = watt-peak.
a. For smaller SHS units only (below 30 Wp capacity).

Table 7.2 Loan Tenure and Interest Rate for POs over Time
Percent

| | IDCOL program year | | | | | |
Refinancing element	*2003–08*	*2009*	*2010*	*2011*	*2012*	*2013–15*
Loan tenure (years)	10	6–10	6–8	6–8	5–7	5–7
Interest rate	6	6–8	6–8	6–8	6–9	6–9
Portion of loan refinanced	80	80	80	80	70–80	60–80

Source: IDCOL.

system costs net of the capital buy-down grant, and the rest is repaid over several years at a flat interest rate of 12–15 percent. These credit terms help to keep the monthly installments within the affordability range of rural consumers. Upon SHS installation, the POs apply to IDCOL for re-financing support for up to a maximum of 80 percent of the microcredit extended to household buyers. The credit terms for PO refinancing are six-to-eight years at a lower interest rate of 6–9 percent. The smaller POs get longer-term refinancing and a lower interest rate than the larger POs. This arrangement provides the POs with liquidity used to install more systems. When the program first started in 2003, the POs were offered refinancing at 6 percent interest rate over a 10-year period (table 7.2). As the program expanded, the refinancing interest rate was increased, the refinancing tenure decreased, and the refinancing percentage reduced with the goal of transitioning the program toward full commercial financing, while ensuring that it remained affordable to the poorer segments of the population.

The incentive mechanism of IDCOL is geared toward promoting competition in the market. By offering the smaller POs refinancing for a longer tenure at a lower interest rate and an institutional development grant, the market competition is being improved. Today, 49 POs are operating in the market though the market is still dominated by large players, with Grameen Shakti accounting for 60 percent of all installations, followed by the Rural Services Foundation (RSF), at 20 percent.

Market Competitiveness

Results of the BIDS/World Bank 2012 survey show that only 17 percent of the PO branches operating in the 64 SHS treatment villages (28 out of 167) operate in a single area, highlighting the market competitiveness of on-the-ground operations.[1] Most POs have been operating for just two years or less. Staff members typically perform more than one job function, including those of manager, accountant, technician, and field assistant. A good proportion of managers have formal technical training and many have worked previously for other POs.

In terms of sales, Grameen Shakti has had the highest monthly target of about 34 units, followed by RSF with 31, Srizony with 18, and BRAC with only 5. On average, a branch has a monthly target of 24–25 new clients, which is quite ambitious, given that SHS delivery and installation require about three days. About 80–85 percent of potential customers targeted by the POs purchase a SHS. Not surprisingly, the POs tend to target wealthier households with steady incomes, who can better service their debt to the POs.

At the time of delivery and installation, both Grameen Shakti and RSF provide consumers training in basic system operation, safety precautions, and maintenance and repair. Among the other POs surveyed, however, officials from only about one-third of the POs indicated that their branches provide consumer training. Among the clients trained, most are male household members; however, women receive training in two-fifths of cases. Follow-up visits usually occur about a week after delivery and installation; however, only 70 percent are scheduled, while the rest are at the consumer's request. At the time of the scheduled visits, only half of consumers indicated a good understanding of SHS operations.

Areas where the POs have been operating for a long period or with less client interaction tend to experience higher rates of irregular payment and delinquency. Currently, BRAC has the highest rate of irregular payment (30 percent) and delinquency (13 percent). But recently established POs are also experiencing higher delinquency rates, as well as clients voluntarily returning units. Of the 1,752 systems returned, 627 were in areas where branches had been in operation for no more than a year. According to PO branch officials, the main reason clients give for returning their units is the inability to pay back the loan. In addition to household financial constraints, newly established POs face problems related to inferior technology and system components (e.g., charge controller and battery), maintenance and management issues, and, in some cases, interference from local politicians, as well as other POs operating in the same area.

On-the-ground operations are fairly standard. Many rules and procedures are similarly implemented across the POs. Yet growing competition, particularly among the newly established, smaller-capacity POs, is, in some cases, leading to a compromise in quality of service. Given Bangladesh's large potential for scaling up the reach of SHS, it is imperative that issues related to technical quality of system components, PO backstopping services, and customer complaints are addressed in a timely manner to ensure the initiative's long-term sustainability beyond the IDCOL program intervention.

Technical Quality: What Are the Gaps?

Although a SHS runs on free renewable energy, the technology is complex, requiring careful, synchronized operation of all system components for optimal performance (annex 7A). Common problems that result in lost efficiency and higher cost per unit of electricity include improperly sized systems, use of poor-quality system components, and installations done under inappropriate rooftop conditions. Inferior system components need to be replaced more often. Laboratory experiments confirm that the typical overall efficiency of a SHS is 10–12 percent, depending on the quality of the electrical devices used. Among all devices, the photovoltaic (PV) panel is the least efficient, followed by the battery and load; the converter also can contribute to overall efficiency loss. Given these technical complexities, quality assurance is imperative.

An independent technical standards committee sets the standards and approves the eligibility of SHS equipment for IDCOL support. Refinancing and grants are released by IDCOL only after SHS inspection confirming that systems are installed according to approved technical standards. When the program started in 2003, IDCOL inspected 100 percent of the systems before releasing refinancing and grants support. As the program expanded, 100 percent inspection was not possible before releasing funds; however, IDCOL inspectors are in the field year-round for inspecting SHS on a random basis. Problems detected during inspection are reported to the POs, and future refinancing and grants support are withheld until satisfactory resolution of the problems is confirmed. All of these measures work together to create a market of high-quality SHS conforming to technical standards.

Consumer Training and Feedback

The survey of PO branches shows that, over the past year, customer misuse accounted for more than half of prematurely damaged SHS units (figure 7.1). This finding highlights the need to intensify basic consumer training, particularly on battery use, maintenance, and recycling.

Figure 7.1 Main Causes of SHS Damage in Past Year

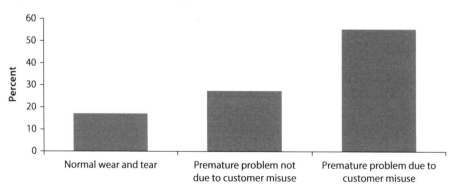

Source: BIDS/World Bank 2012.
Note: SHS = solar home system.

Institutional Capacity and Management

Owing to the tremendous growth in SHS use over the past few years, the POs have had to regularly supply component parts; some may lack sufficient institutional capacity and technical skills to manage these installations.[2] Such issues may have contributed to problems the POs have recently reported regarding inferior-quality components. The survey results show that most POs face problems related to poor-quality lights and charge controllers, while a significant number also have battery-related problems (figure 7.2).

The POs have programs in place to inform buyers how to safely dispose of damaged components, including batteries and charge controllers; however, the study survey findings show that more than half of consumers either dispose of components in an unsafe manner or recycle them for activities that are hazardous to the health of users and the environment (figure 7.3).

IDCOL's recent efforts to strengthen battery recycling have made it mandatory for all battery manufacturers and recyclers to have ISO 14001:2004 and

Figure 7.2 SHS Component Problems Reported in Past Year

Source: BIDS/World Bank 2012.
Note: SHS = solar home system.

Figure 7.3 Client Distribution for Returned Batteries

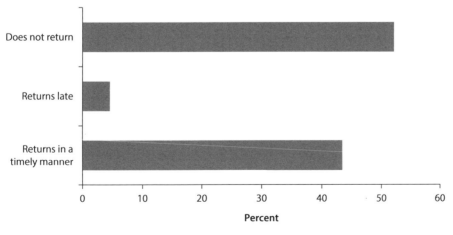

Source: BIDS/World Bank 2012.
Note: PV = photovoltaic; SHS = solar home system.

OHSAS 18001:2007 certifications to be eligible for the SHS program. These requirements have resulted in all local battery manufacturers and recyclers being in compliance with international standards for safe disposal and recycling of batteries. In addition, IDCOL has increased incentives to households and POs for the return and collection of expired batteries to ensure they are returned to the approved recycling centers and not backyard smelters. It is too early to report on the results of these recently introduced incentive measures, and continuous monitoring by IDCOL will be needed to ensure that expired batteries are collected and recycled in an environmentally safe manner. There is also a need for greater consumer training to avoid system misuse and ensure the safe handling of damaged components.

The household clients' problems after delivery and installation have varied (figure 7.4). Of the 1,600 consumer households, only about 1 percent (18 customers) have reported problems that required PV panel replacement. By contrast, about 18 percent (287 customers) have had to replace damaged or broken charge controllers, while nearly 11 percent (170 customers) have had to replace batteries. Premature replacement of both charge controllers and batteries indicate that some components have been of low quality or operated improperly.

Replacement patterns show a significantly higher reliability for compact fluorescent lamps (CFLs) than for tube lights. During the 12-month period prior to the survey, more than half of consumers had to change their tube lights, compared to only 14 percent that had to change their CFLs (figure 7.5).

Figure 7.4 Problems Faced by Client Households Since SHS Installation

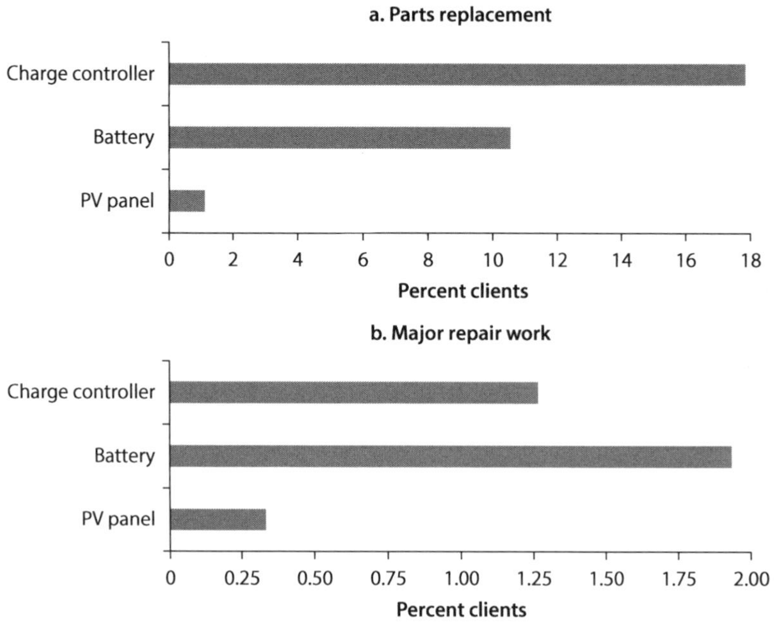

a. Parts replacement

Charge controller, Battery, PV panel — *Percent clients* (axis 0 to 18)

b. Major repair work

Charge controller, Battery, PV panel — *Percent clients* (axis 0 to 2.00)

Source: BIDS/World Bank 2012.
Note: PV = photovoltaic; SHS = solar home system.

Figure 7.5 Change of Tube Lights and CFLs in Past Year

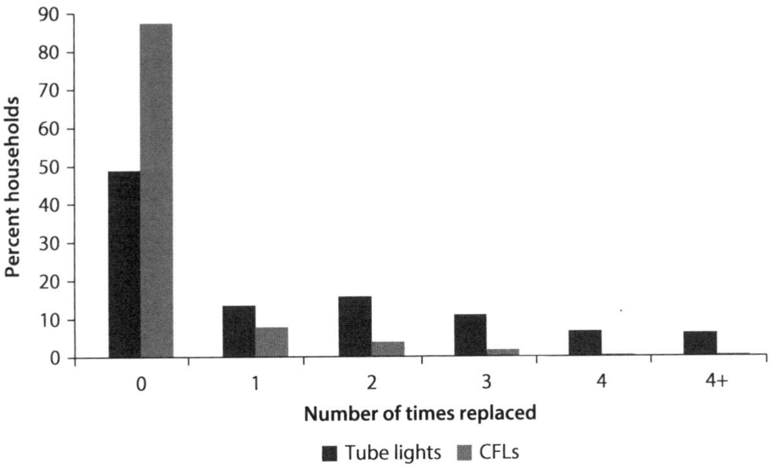

Source: BIDS/World Bank 2012.
Note: CFL = compact fluorescent lamp.

It is unclear whether these replacements were within the warranty period or were for the same connection points. Also, clients could have been unaware of technical compatibility issues, which resulted in damage, suggesting the need for IDCOL to set and enforce more rigorous regulations.

Standards-Testing and Design

Numerous local and foreign manufacturers are involved in providing SHS components of varying quality in Bangladesh. As the pace of SHS installations has accelerated, concerns about quality and system efficiency have increased. Standards-testing for quality is needed to increase the life cycle of the SHS units, reduce environmental damage from improperly discarded components, and increase the feasibility of a future scaled-up program (Khan, Rahman, and Azad 2012). Recent efforts by IDCOL to work with the Bangladesh University of Engineering and Technology (BUET) for establishing a PV testing facility in Bangladesh is a step in the right direction.

The quality of PV panels and other imported components is assured through international standards; however, no similar standards-testing is available for locally assembled PV panels. Because of their cheaper cost, along with consumers' lack of knowledge, PV modules with poorer fill factor and efficiency are flooding the Bangladesh market outside the IDCOL program. It is suggested that IDCOL, working through its technical standards committee, set standards to push local assemblers to improve their products in order to develop a local market for quality PV panels. Although not needed immediately, an appropriate mechanism for the safe disposal of PV panels—assuming a 20-year life cycle—will need to be introduced at some future point to ensure program sustainability.

Public Awareness Building

It is suggested that government and the private sector work together to raise public awareness about solar PV and other renewable energy technologies; this, in turn, would increase SHS popularity and thus contribute to building rural market demand. IDCOL's technical standards committee could organize special training programs on the effective use of SHS, while subsidized research could focus on increasing the efficiency of SHS components and the overall system. To ensure effective research and development on these issues, links should be developed between authorities, industries, and universities.

Demand Dynamics and Market Size

Various intervening demand-side factors are likely to interact with the technical quality issues discussed above to influence the growth potential of the rural SHS market. These include raising household incomes, need for higher levels of energy over time, and potential expansion of the rural grid system in currently off-grid areas. Emerging new technologies would make it possible for a cluster of households to be efficiently served with centrally located panels and batteries rather than individual SHS (e.g., mini- or micro-grids). All such factors need to be taken into account when estimating the future potential of SHS in Bangladesh.

As previously mentioned, the POs tend to target creditworthy clients to ensure loan repayment. Once the needs of wealthier off-grid households have been met, the POs may consider an area saturated and turn to meeting the needs of grid-connected households that desire SHS as a backup power supply and can pay at the unsubsidized cost.

Recent technology improvements suggest that it should be possible for the POs to expand their market reach to include more socioeconomically disadvantaged groups, particularly women. Because of energy-efficient light-emitting diode (LED) technologies, a cheaper 20 Wp system can satisfy the same lighting needs that would have earlier required a higher-priced, 40 Wp system. Since the introduction of LED technologies in early 2011, the 20 Wp system has been the most sold unit. Currently, 62 percent of the systems sold each month are 20 Wp units, compared to 11 percent that are 65 Wp units. Under the prevailing system of incentives and prices, without grid expansion in SHS areas and without SHS adoption by grid-connected households, the simulated market size for SHS is about 8 million, as shown in chapter 6.

Concluding Remarks

IDCOL's efforts at transitioning the program toward commercial financing without the capital buy-down grant need to be carefully reviewed in light of the high cost of financing in the local market. With enhanced competition in the market and technological improvements, SHS prices may decline in the future. However, with the relatively better-off households already reached, future program expansion is likely to reach relatively poorer segments of the population. Too abrupt a withdrawal of capital buy-down grants or concessional financing may result in the SHS being too costly for the poorer segments. In the absence of a formal regulatory authority in place, IDCOL's role in setting and enforcing standards will be critical to ensure development of a market for quality SHS.

This chapter has highlighted the technical/quality issues that need to be addressed on the supply side to ensure the successful scaling up of the rural SHS market. It has also highlighted how technology improvements can affect consumer demand as the market continues to develop. Even as new technologies make further inroads and the grid expands its reach, there is a large potential for expanding the SHS market, particularly for lower watt-peak models. The benefits that have already accrued to rural households from 3 million SHS units installed to date underscore the urgent need to address the technical-quality issues with a view to developing a sustainable market going forward.

Annex 7A: Analysis of SHS Technical Efficiency

The overall technical efficiency of a solar home system (SHS), illustrated in figure 7A.1, is equivalent to the product of the efficiencies of the individual components.

This is expressed by the following equations:

$$\text{Overall efficiency} = G_{pv} \times G_{converter} \times G_{charge\ controller} \times G_{battery} \times G_{lamp}$$
$$= 0.16 \times 0.95 \times 0.955 \times 0.9 \times 0.845 = 0.1098$$

$$\text{Output power} = \text{Overall efficiency} \times \text{Input power} = 0.1098 \times 1{,}000\ \text{W/m}^2$$
$$= 109.80\ \text{W/m}^2.$$

If the DC-to-DC converter can be removed from the system, overall efficiency of the SHS increases.

Figure 7A.1 SHS Block Diagram, Including DC-to-DC Converter

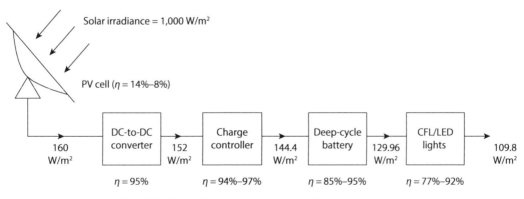

Note: CFL = compact fluorescent lamp; LED = light-emitting diode; PV = photovoltaic; W/m² = watts per square meter.

Notes

1. These are distributed as follows: Grameen Shakti, 12; BRAC, 1; Srizony, 4; RSF, 2; and other POs, 9.

2. An earlier KfW report that monitored the technical quality of installations underscores the need for the POs to increase their technical management capability and use of skilled manpower.

References

BIDS/World Bank. 2012. "Household Survey Data on Impact Evaluation of Solar Home Systems in Bangladesh." Bangladesh Institute of Development Studies and World Bank, Dhaka.

Khan, S. A., R. Rahman, and A. Azad. 2012. "Solar Home System Components Qualification Testing Procedure and Its Effects in Bangladesh Perspective." In Proceedings of IEEE Global Humanitarian Technology Conference: Technology for the Benefit of Humanity, Seattle, October 21–24.

CHAPTER 8

Overview, Policy Perspectives, and Emerging Issues

Achieving universal electrification in Bangladesh is impeded not only by lack of income but also by the country's inability to generate and distribute enough grid electricity to meet demand. The grid-based electrification in Bangladesh has reached only about two-fifths of rural households. The high cost of grid electricity in remote villages means that power utilities concerned about their financial viability may prioritize large industrial loads in urban areas over the rural countryside, where the vast majority of Bangladeshis live. Even households connected to the national grid experience frequent and prolonged power outages owing to limited electricity generation and supply. Electricity generation constraints are forcing utilities to slow down grid extension in rural areas. Universal access to electricity by 2021 is a stated goal of Bangladesh's national strategy; but expecting to achieve it through reliance on grid electrification alone would be unrealistic. In this context, solar power using photovoltaic technology, known as solar home systems (SHSs), is a promising instrument for promoting electrification in remote rural areas that would not otherwise receive a grid connection in the foreseeable future. This book evaluated Bangladesh's remarkable growth in SHS growth and its support schemes, as well as welfare impacts using both household survey and institutional data.

Surge in SHS Growth in Off-Grid Areas

Our analysis clearly shows that households in off-grid rural Bangladesh are turning to SHS as a viable alternative to conventional power supply. In fact, Bangladesh has witnessed the world's fastest growth in off-grid SHS coverage. Installations have accelerated from just 15,745 installations in 2003 to 3 million today. What started in 2003 as a World Bank–funded project with a target of 50,000 SHS installations over a five-year period is now installing more than 50,000 systems a month with support from the World Bank and other development partners. Despite recent phenomenal growth, only 10 percent of off-grid rural households had been reached by early 2014, suggesting ample room for continued expansion. The Infrastructure Development Company Limited (IDCOL), the project's implementing agency, has set a target of installing another 3 million SHS units within the next two years.

SHS offers households a convenient supply of electricity for lighting and running small appliances for about 3–5 hours a day. The main components of the SHS are the solar panel—the heart of the system—charge controller, and rechargeable battery. Usually installed on the rooftop of a house at an angle designed to collect maximum sunlight, the solar panel converts sunlight into electrical energy. The capacity range for most units installed in Bangladesh is 20–120 watt-peak (Wp). A unit with 50 Wp capacity can power four lights, a mobile phone charger, and a television set (including a color one).

Study Overview

This study assessed the welfare impact of SHS on households in rural Bangladesh and the institutional structure and financing mechanism, including subsidies, currently in place. Underlying this broad goal is the recognition that households want cheaper systems and quality service, while suppliers require a reasonable market-based profit to stay in business. The study entailed an intensive empirical investigation. Primary data consisted mainly of a large-scale, nationally representative household survey conducted in 2012. Surveys of IDCOL's nongovernmental organization partner organizations (POs) and communities were also conducted. The research issues analyzed were grouped according to general and gendered household impact, program delivery and monitoring of technical standards, market size and demand, and carbon emissions reduction.

The IDCOL-administered SHS program is based on a well-designed, effective structure, comprising various entities. IDCOL's effective microcredit-based delivery mechanism also benefits household consumers and the POs. Consumers can purchase the SHS units from the POs on credit for two-to-three years at a flat interest rate of 6–12 percent after making a down payment of 10–15 percent of the system cost. Among the POs, Grameen Shakti—a subsidiary of the Grameen Bank, which has a long history in microcredit lending and an extensive countrywide network—is the largest, currently accounting for 60 percent of all installations, followed by Rural Services Foundation (RSF), at 20 percent. IDCOL's grant and subsidized loan policies have been instrumental in pushing the solar frontier by shifting demand through the entry of smaller POs that market a range of panels. The capital buy-down grants, which were US$70 per system when the program first started in 2003, have come down gradually to US$20 (for smaller systems only). The declining rates of grants raise the unit cost of operation. But increasing PO competition and growing market demand, driven, in part, by advances in solar panel technology, have resulted in the decline of unit prices despite the subsidy reduction. Interestingly, the subsidy has declined at a higher rate than the unit price.

Drivers of SHS Adoption and Benefits to Households

The study found that household wealth, education of household members, particularly women, and pricing are key drivers of SHS adoption in off-grid villages. An examination of SHS adoption rates by landholding, a proxy for household wealth, showed that only 10 percent of households with low-to-medium landholdings had adopted SHS, while the rate had surged for large landholders. The findings also showed that wealthier households demand higher-capacity units, and women-headed households are more likely to adopt them. SHS adoption changes the composition of household energy consumption, significantly reducing the carbon emissions from fossil-fuel consumption. For example, by late 2012, SHS adoption had resulted in savings of more than 40 million liters of kerosene. The direct emissions reduction amounted to more than 240,000 metric tons of carbon dioxide (CO_2). When a household first adopts a SHS, the immediate benefit is the replacement of polluting kerosene lamps that provide the dimmest of lighting with non-polluting, high-quality solar lights. Better lighting has the immediate effect of extending the waking and working hours of family members, allowing for school-going children and adults alike to study and read in the evenings. There is also a greater sense of security at nighttime, allowing for greater social interaction.

With better-quality kitchen lighting, women spend less time cleaning and can cook more efficiently in the evenings, which has nutritional benefits for family members. With access to solar-powered lighting and television, family members' time-use patterns change in ways that lead to longer-term socioeconomic benefits. With better-quality lighting, school-going children—both girls and boys—study longer in the evening, which can positively impact educational outcomes. Women with access to higher-quality kitchen lighting can manage their household chores at a less hurried pace throughout the day and redirect their freed-up time to income-generating, educational, and leisure activities.

Ultimately, solar-powered electricity facilitates a virtuous cycle of growth in household consumption and income. First, through reduced kerosene consumption, households save money that can be redirected to purchase food and non-food items. Second, changes in time-use patterns and decision-making power encourage the reallocation of household resources for consumption and production, which, in turn, increases both income and consumption. Encouraging the promotion of home-based businesses, such as charging mobile phones, directly affects income and thus indirectly impacts consumption. With longer exposure to solar power, the amount of electricity consumed increases; and growth in food, non-food, and total per capita expenditure accrue over time. The study found that the accrued benefit of a SHS unit exceeds its cost by 500 percent, meaning that SHS adoption is quite cost-effective for households.

Subsidy Effectiveness and Emerging Issues

Given the recent surge in SHS adoption in off-grid areas under the refinancing scheme introduced through IDCOL, a key policy issue is whether or how much of the subsidy is worth supporting. The analysis shows that the subsidy has declined over time, from 25 percent of the average unit price in 2004 to 10 percent in 2012. However, as the demand for a SHS unit is price-inelastic, the POs could charge a higher price for a SHS at cost, perhaps without a substantial reduction in market demand. But poorer households, who may prefer less-expensive, lower-capacity units, might not be able to afford the market price. This suggests that IDCOL's subsidized marketing operation, even at less than 10 percent of the current unit price, might be needed for some time.

That said, much can be done to improve on-the-ground operations to ensure the sustainability of future market expansion. Given the highly competitive nature of the environment in which the POs operate, there is an urgent need to balance faster delivery and installation of SHS units with a stronger focus on quality of service. Key areas to address include ensuring the technical quality of installations; enforcing regulations on standards and specifications for system components; improving the targeting of after-sales service and training, particularly among women-headed households; and taking actions based on consumer feedback. Tackling these technical, managerial, and operational issues at this stage of market development can create a win-win situation: households can receive affordable quality products and services, and the POs can maintain a reasonable profit to sustain their market operation.